ENTERTAINMENT TECHNOLOGY PRESS

In taking advantage of the latest in 'print on demand' digital printing techniques, Entertainment Technology Press is approaching book publishing in a very different way. By establishing a wide range of highly specific technical books that can be kept up-to-date in a continuing publishing process, our plan is to cover the entertainment technology sector with a wide range of individual titles.

As will be seen by examining the back cover of this book, the ETP list is divided into various categories so as to allow sufficient room for generic growth and development of each title. To ensure the quality of the books and the success of the projec[illegible] building up a team of authors who are practi[illegible] by the industry. As specialists within a pa[illegible] anticipated that each author will stay cl[illegible] title or titles for an extended period.

All Entertainment Technology Press tit[illegible] on the publisher's own website at www.[illegible] where latest information and up-dates can be accessed by purchasers of the books concerned. This additional service is included within the purchase price of all titles.

Readers and prospective authors are invited to submit any ideas and comments they may have on the Entertainment Technology Press series to the Series Editor by email to editor@etnow.com.

Entertainment Technology Press Ltd
The Studio, High Green, Great Shelford, Cambridge CB2 5EB
Tel: +44 (0)1223 550805 Fax: +44 (0)1223 550806

HEALTH AND SAFETY MANAGEMENT IN THE LIVE MUSIC AND EVENTS INDUSTRY

Chris Hannam

ENTERTAINMENT TECHNOLOGY PRESS

Safety Series

This book is dedicated to my late friend, Keith Ferguson.

HEALTH AND SAFETY MANAGEMENT IN THE LIVE MUSIC AND EVENTS INDUSTRY

Chris Hannam

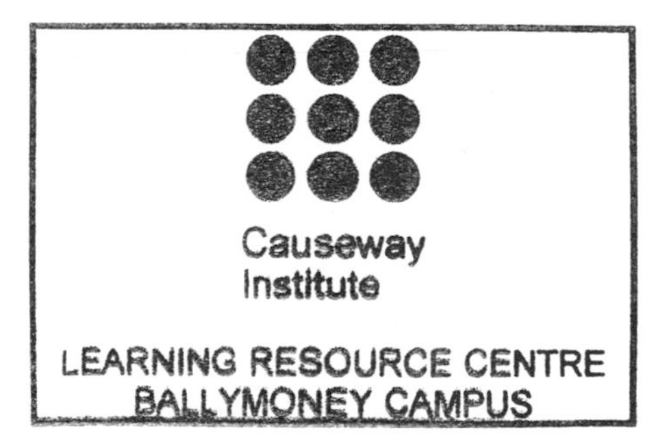

Entertainment Technology Press

Health and Safety Management in the Live Music and Events Industry

First edition Published October 2004 by
Entertainment Technology Press Ltd
The Studio, High Green, Great Shelford, Cabridge CB2 5EG
Internet: www.etnow.com

ISBN 1 904031 30 7

A title within the
Entertainment Technology Press Safety Series
Series editor: John Offord

CODE / HSLM001

CONTENTS

ACKNOWLEDGEMENTS

I would like to thank the many and various friends and colleagues who supplied (either wittingly or unwittingly) the inspiration, ideas, information, incentive, comment, criticism and knowledge that made this book 'happen', for they have to take some of the blame!

They are: Neil McLaren, Billy Carr, Bill Egan, Nick Cooke, Kim Ray, Pete Howard, Jon (JC) Corblishly, Mick Upton, John Jones, Richard Stevenson, Julian Walden, Andy Lenthall and the person who originally got me interested in health and safety, Bill Goodall. Sorry about this, Bill, but you really do have a lot to answer for; you must now face the consequences and take a major part of the blame.

The section on Safety Bonds was written by Chris Higgs, who also made major contributions to the Work at Height and LOLER chapters – so an extra special mention must be made to him. Cheers, Chris!

I must also acknowledge the HSE, whose books and publications (together with the odd leaflet and publication from the ABTT and the British Safety Council) I have plundered for much of this publication's content.

1 INTRODUCTION

I am sure that we all have often heard grumbles along the lines of: "I wish those Health and Safety people would get off our backs. If they had any idea at all, they would know we have a very good safety record in our industry, so we don't need them to tell us how to do our jobs safely".

As an answer to this I would like to quote Tim Norman of Edwin Shirley Staging who so rightly states:

> "We have a very good record for Health and Safety within our industry, but the reason we have such a good record is not necessarily because we are safe, but because there are *no* records!"

Records aside, it's a sad fact that very few companies or individuals in our industry have any idea of the law and regulations that govern us on Health and Safety, thereby leaving themselves, and others, wide open to financial loss, civil claims and prosecution for negligence and injury. On the other hand, there are many who *think* they know all they need to know but many of their ideas are misguided or distorted and don't offer any real protection. This is a risk none of us can ever afford to take.

These statements appeared in the first edition of this book and in some respects a lot has changed since then as a few more regulations have been introduced and many people are more aware of what is expected of them. But in other respects nothing much at all has changed. Despite being aware and knowing what is required of them, the majority still do very little about complying with H&S law and regulations, thus putting themselves and others at risk unnecessarily. Others produce reams of wonderful health and safety documentation such as Risk Assessments and Health and Safety Policies, but then do absolutely nothing about implementation and operating the required control systems. We still have a long way to go when it comes to getting our industry house in order in terms of health and safety. This book is an attempt to draw to people's attention the kind of thing they really *should* be doing and, with any luck, an added bonus will be to drive out some of the 'cowboy' companies that bring the industry into disrepute.

Another reason for not having any records in our industry is the large number of self-employed people who, if injured at work, simply limp away into the sunset without making a fuss in case they are classed as troublemakers. They feel being classed as such will affect their chances of obtaining future work, this being based on the natural assumption that no one wants to employ a troublemaker. One further reason is that most people who join our industry did not do so "to get into health and safety", they joined because they were a kind of rebel or different in some way - a bit like running away to join the circus.

This book only touches on some of the regulations covered by the Health and Safety at Work etc Act 1974 (HASAWA) and is by no means exhaustive. The intention is to map out the territory ahead and get you started on the road to good health and safety management.

I have often heard it said that health and safety is just common sense. If that is the case then why are there still accidents taking place? Why are people still designing arenas for outdoor events badly? Why are 'pit' areas still being designed badly? Why are promoters allowing bands on stage that incite dangerous crowd behaviour?

(Of course, I am assuming that we all have a degree of common sense. But then again, and as any good production manager will tell you: "assumption is the mother of all cock-ups", so perhaps I shouldn't make such a rash assumption).

Under the pressure of making a show 'happen', health and safety is very often overlooked because it was not included in the planning stage or within the budgets for the operation and therefore no systems were put in place to control safety. One aspect of this book is to demonstrate what the law expects us to do in terms of implementing control systems and how the law expects us to demonstrate, usually with documented evidence, that we have actually done what is expected of us in this respect. If things do go seriously wrong, however, how are you going to prove that you had done everything that could reasonably be expected of you to have prevented that accident taking place? It must always be remembered that documentation alone is not sufficient. Many companies now in our industry seem to have piles of health and safety documents but do not implement them on site. They are as much cowboys as those with no documents. The objective is to have all the right documentation that is also in operational practice.

H&S is an essential part of quality control. A business with good health and safety controls in place is in all probability a well-run 'quality' business. Unfortunately, H&S management is usually the last consideration for many poorly-run outfits and doesn't even register with the kind of people who arrive on horseback and wear big hats. As a 'rule of thumb' guide, if H&S is looked after, the rest of the business is usually well run and in order. Quality and H&S go together hand in hand; they are different sides of the same coin.

With a better understanding of H&S Management it will become very obvious as to how quality management and H&S Management are interlinked. H&S is a management responsibility of equal importance to productivity and quality.

Effective H&S management is not 'common sense', rather it is based on a common understanding of risks and how to control them. This can only be brought about through good management thus leading to a cooperative effort at all levels in an organisation.

In the last five or six years I have seen changes starting to take place regarding people's attitude towards health and safety, together with a growing number of companies really starting to make very serious efforts towards better health and safety management.

Quite surprisingly, the pattern that seems to be emerging is that it is not the large, well-established companies that are making the first moves and setting the standards, as one would expect, but the small and medium-sized operations that are keen and eager to establish themselves.

These new and forward-thinking businesses can see that having the necessary health and safety management systems in place is the way forward to establish themselves.

They also see that having health and safety management systems in place as being the way towards gaining preferred or approved contractor status and thus getting a chance to tender for those vital but otherwise out of reach contracts.

The companies that are not "getting their act together" so well in terms of health and safety are the large well-established companies which seem to have the attitude that no one knows better than they do and that they have the best crews who know it all anyway!

Sorry guys, but how wrong you are! You may have the best LDs, the best gear, the best sound engineers and the best riggers, but most of them have no idea about health and safety and the way it should be managed. Most of them 'think they know', but the truth is very different. Even if they have been doing the job for twenty years, they are a long way off target and such an arrogant attitude will get you nowhere.

Of course there are a good number of companies that are new and far less established which still need to get their houses in order in terms of health and safety.

This is the expected situation with these small companies as they generally lack the vital experience, contacts and information exchange links required to gain the information that is needed to put the required health and safety management systems in place.

But We Already Have The "Pop Code" To Tell Us All About Health And Safety

Until a few years ago I (no doubt like many others) thought that if the Entertainment Licence conditions and the guidance set out in the original 'Pop Code' were followed, then that was all we needed to do. *If only it were that easy.*

If that was the case then we have already overlooked the implicit legal requirement to comply with the HASAWA and the various regulations it covers.

Even though it *did* mention the requirement to comply with the HASAWA in the old Pop Code *and* in some Entertainment Licence conditions, it generally got overlooked. Worse still, it was often considered that the Pop Code was the same as HASAWA and therefore we need do no more to meet our legal obligations. I now know, and after reading this you should realise, that we have a lot more to do to put our house in order.

It's still no surprise to find that of all the suppliers I am required to book for events and major festivals, very few of them understand their basic responsibilities under the HASAWA, the various regulations it covers, or what is legally required of them.

These days there is a lot more to Health and Safety than just seeing that a safety rail is fitted to a stage.

In most cases, the old Pop Code set out the basic standards for Health, Safety and Welfare that should have been achieved for the people attending the festivals, concerts, shows and gigs we produced. The finished product, in other words.

The Pop Code is only guidance to help us meet the conditions included in an Entertainment Licence; it was not the same as the HASAWA which set the basic standards of H&S that we *must achieve* during *our* shows (including the set ups and breakdowns).

The HASAWA is designed for *our benefit* to maintain *our own Health, Safety and Welfare* as well as that of the public. The major difference is the

Pop Code is only for *guidance*, the HASAWA is *law.*

The current edition of the Pop Code (The Event Safety Guide, A Guide to Health, Safety And Welfare at Music and Similar Events, also known as the Purple Guide or Son of Pop Code) varies considerably from the first edition in that it is far less prescriptive, that is to say it has fewer 'dos and don'ts' and does not stipulate so much the items or number of items that must be in place.

The reason for this is that all health and safety legislation and regulations are being de-regulated to become less prescriptive, and instead the risk assessment process is applied to work out many of the requirements; a good example of this being shown in the Health and Safety (First Aid) Regulations 1991. At one time the regulations gave figures on how many First Aiders were required in a given situation. Now you are required to work it out for yourself after risk assessment. The other major difference with the new Event Safety Guide is that it now looks a little harder at the HASAWA and the various regulations under it, but unfortunately not in much detail. This book, I hope, will make a good companion to the Event Safety Guide and will help you understand the regulations a little more, but it's not a substitute to the Event Safety Guide or a guide to crowd safety and welfare. A guide to crowd safety and welfare would take another book.

Where the Event Safety Guide is really effective is as a source of reference material, and a useful checklist when planning an event and when carrying out risk assessments. It is also used by local authorities when considering appropriate Entertainment Licence conditions and restrictions.

I know a few people who think that the Pop Code should be law and not just guidance, but that's not possible; it is only a guide to what we should consider warrants a risk assessment. If it were law we would have to apply the same control systems to a classical concert as we do at a major rock festival for 100,000 people!

I Don't Need All This Health And Safety Stuff, I've Got Insurance!

Believe it or not I was actually quoted the above words by one company I used to deal with. The answer to that is that if there is a claim on your insurance, the insurance company will always pay out because they have a contract with you to do so and they have to protect any (innocent) third parties.

But if it is found that you have been negligent when it comes to health and safety you will be sued by your insurance company to claim every penny (plus costs) back from you personally, and while you may be able to insure against civil claims you can't insure against a criminal prosecution, and failing to comply with the HASAWA is a *criminal offence.*

It's a little like finding your car insurance is invalid after an accident because you don't have an MOT certificate. Insurance is not a 'Get Out Of Jail Free' card, as there is a good possibility that deep within the small print of your insurance policy is a condition that requires you to comply with the HASAWA and all the regulations under it, and even if it's not actually stated, you still have a legal obligation to comply.

In a way, good health and safety management is insurance. By putting management systems in place you are not only preventing accidents but you are also protecting yourself from prosecution and associated losses.

So with this in mind, please read on, but remember: Health and Safety is a very big subject and this is just a very small book.

The first edition of this book was called 'An Introduction to......' and I spectacularly managed to get a lot of things wrong in that first edition which I have now corrected. My own knowledge of the subject has also grown in the interim so I have included a lot of that new information, but I may have to go to a third edition at the rate legislation is changing. It's an endless job trying to keep up with the law.

At this point I must be totally honest and say that most of this book was not written by me at all, it has been 'compiled' from snippets (and big chunks) originating (and stolen) from various HSE publications, quotes, comments, conversations and articles that were produced by people far more intelligent than me (and besides that I don't want to reinvent the wheel). So, I acknowledge you all (especially the HSE) as the original authors of all the work I have copied here.

FACT

There are over 400,000 reports made under the Reporting of Incidents, Diseases and Dangerous Occurrence Regulations each year.

256 of these are fatal
30,000 are major injuries
134,000 involve people being away from their *normal* job for at least 3 days
10,000 are dangerous occurrences.

PREMISE

75% of all accidents are foreseeable and preventable.

2 WHAT IS MANAGEMENT?

The definition is: *"The effective use of resources in pursuit of organisational goals"*.

'Effective' implies achieving a balance between the risk of being in business and the cost of eliminating or reducing those risks. Management entails leadership, authority and coordination of resources.

Health & Safety Management is no different from any other form of management. It covers:

a) The management of the Health & Safety operations at national and local level – setting of policy and objectives, organising, controlling, and establishing accountability.
b) Measurement of Health & Safety performance on the part of the individuals and specific locations.
c) Motivating managers to improve standards of Health & Safety performance in those areas under their control.

Health & Safety is about *your* health, *your* safety and *your* welfare. It is not about stopping you from working or interfering in your work, it's about helping you to work, keeping your liberty, saving you money and protecting your business.

As we all know, we have a legal duty to keep detailed records, have audits carried out and submit income tax and VAT returns for the financial aspects of our business. The safety side of the business is no different – we have similar legal requirements to keep detailed records, provide information, have safety audits carried out and to submit reports and returns. The essential difference is that strict enforcement is carried out on the financial side of business, the reason being that the government wants your money. This is why companies employ accountants to ensure the requirements are met, to reduce liabilities to a minimum and avoid prosecution. Although the responsibilities are just as great on the health and safety side and businesses are required to appoint safety advisers to help them reduce their liabilities and avoid prosecution, enforcement by the government is not so rigorous. The reason for this is simple: there is little or no money in it for them; it is not a revenue raising priority.

Having said that, severe penalties exist should you fall foul of health and safety law.

It should be fairly obvious, even to the half-brained, that quality Health & Safety Management plays a very significant role as a financial loss control system within a business operation and also plays a part as an important marketing tool.

The other thing Health & Safety provides is a weapon. In fact it's the only weapon the HSE and the trade unions have against the cowboy operators who undercut the good suppliers and then provide unsafe and poor quality products – not that I am suggesting we become trade unionists.

So if you or your company don't want to be classed as cowboys, then I suggest you put your horse in the stable, hang up your guns and start putting your house in order in terms of health and safety and join the fight against them. But make sure your own house is in order first, otherwise you are just as much a cowboy as they are.

In my opinion there are very few real health and safety problems. The great majority of the 'supposed' problems that come to light are due to a lack of knowledge and/or understanding of existing statutory requirements and of an even greater reluctance to put into place health and safety management systems that are combined with the effective and correct means to implement such systems.

This is a very shortsighted attitude that, if not changed, may cost far more than full implementation and compliance with Health & Safety Law.

Many people also think there are a lot of new regulations being introduced, but this is not really the case. The truth is that there are a number of old regulations that (in these days of litigation) people are now becoming more aware of.

3 THE LICENSING ACT

This new Act (which comes into full force in 2005) is an attempt to amalgamate five other different licensing regimes with public entertainment licensing, including alcohol, cinemas, theatres, late night refreshment houses and night cafés. The Act replaces 22 existing Acts of Parliament and involves changes to over 60 others; some existing legislation will still apply in Scotland. In this chapter I have intentionally only looked at the Entertainment Licensing aspect of the new Act and not paid too much attention to alcohol licensing.

The Act seems to have extensively tackled the subject of alcohol but entertainment licensing appears to have been 'tacked on the end' as an afterthought. A number of 'grey areas' still exist in the department of entertainment licensing that obviously need to be tackled. Under the Act all catering and food concessions (including mobile units) that operate late at night (between 11pm and 5am) will now require a Licence!

Under the new Act an organiser will be required to obtain a 'Premises Licence' from the local authority instead of a Public Entertainment Licence. The 'Premises Licence' will specify the operating conditions for an event where there might be an impact on crime and disorder, public safety and nuisance factors. The conditions that may be set out in a Licence are *in addition* to other statutory health, safety and welfare regulations. In other words, there is a legal requirement to comply with the Health and Safety at Work Act 1974 and the statutory regulations under this Act even though they may not be mentioned in any Licence conditions; this is what is known as an 'implicit' duty. In addition to controlling Health and Safety, the four main objectives of the Licensing Act are:

1) To prevent crime and disorder
2) Prevention of public nuisance
3) To maintain public safety
4) Prevention of harm to children

The Act has made provision for temporary or occasional events by means of a 'temporary event notice' (TEN) for events for fewer than 500 people and lasting less than 4 days (Consecutive Events running one immediately after the other are disallowed) whereby, unless there are exceptional circumstances, the organiser will simply be required to notify the police and licensing authority by means of a TEN at least 10 working days before the proposed event and pay a set fee of about £30. The police can object to a TEN but only on the grounds of prevention of crime and disorder.

Events for over 500 people will require a Premises Licence.

Temporary Licences will be available to individuals who are not licence holders at named premises for a maximum of five events per year, and to holders of Personal Licences for up to fifty events per year, regardless of whether alcohol is sold or not.

Regulations will be made by the Secretary of State to advertise an application for a Licence; this may include notification to residents within a certain radius of a venue, publication of notices in newspapers, etc.

One aim of the Act is to bring consistency and clarity to licensing fees. At present, a licence fee for an event for 5000 people can be between £50 and £5000 depending on the local authority. Charges for a Premises Licence will be set centrally by the Secretary of State, but it is unclear at this time how the charges will be worked out, particularly for outdoor events.

Local Authority Licensing and Environmental Health Officers (EHOs) will be responsible for Health and Safety and licensing enforcement, but they are *not* to be confused with Health & Safety Executive Officers. Licensing Officers and EHOs carry out inspections to ensure that the health, safety and welfare regulations and the conditions of the licence are being complied with. The local authority will be responsible for prosecutions for any breaches of a licence or licence conditions.

The local authority has to grant a licence without a hearing unless representations/objections are raised against a proposal, in which case the Council's Licensing Committee or delegated panel will determine the application. The Licensing Committee or panel will consist of elected members of the Council and sub committees (panels) who will be charged with the determination of applications on behalf of the committee.

Local residents, resident and business associations may make representations to the council against any applications which will then result in a hearing unless the objections are irrelevant, vexatious or repetitive.

The government is expected to publish guidelines on how committees are to be run. This will include how often meetings are to be held and what sort of evidence can be presented to the committee.

The council will be required to consult on its licensing policy, and the policy then published and reviewed at least every three years. Consultations in respect of the policy will be with the emergency services, trade and resident representatives.

A local authority will consider the details of an application. In this process, the authority can alter the terms applied for if any objections are received or the general objectives of their licensing policy are contravened. If no objections are received and the view is taken that no alterations are required to promote the licensing objectives, then it must grant the licence by the terms sought by the applicant. Conditions may also be set on any licence that may be granted. During the application process, the local authority will also consult with the police and emergency services to whom notification of the application must be supplied. They may also inform the HSE in some cases. The police and emergency services will make major contributions towards the conditions on any licence that may be granted.

The police and emergency services may or may not have a direct presence at an event but they have a responsibility to attend any emergencies. The local authority, police and emergency services will probably make a charge for their services at large events with a temporary licence, such as a festival.

The conditions attached to a licence may include such matters as:

- Start and finish times
- Noise restrictions
- Electrical and structural safety
- Adequate access for emergency vehicles
- Ticket sales and audience capacity
- Traders and concessions
- Litter and waste disposal
- Toilets, water and welfare facilities
- Stewarding, numbers and details of training
- Traffic management plans
- Restricted and banned items or practices
- Fire safety conditions and medical facilities
- Emergency and contingency plans
- Alcohol sales

This is not an exhaustive list.

As mentioned, existing Health and Safety legislation will still apply but there is an important new implication for venues: the local authority and fire brigade will no longer be certifying venues. It will be the venue's responsibility to ensure that all work meets the appropriate required standards (Building Regulations, Fire Regulations, Noise at Work Regulations, Electrical Installation Regulations and Health and Safety Regulations etc.). Some venue operators are anxious that their insurance will rise or become unobtainable as a result of the 'self assessment' of safety matters.

It is a criminal offence to organise public entertainment without a licence or to be in breach of licence conditions. In a Magistrates' Court each breach of conditions currently carries a penalty of a £20,000 fine.

Before we start complaining about the refusal to grant a licence, it is worth remembering that a Local Authority or individual officers of an authority can be held negligent if they do not ensure that an event is 'safe' when they grant a licence. A charge of negligence could result in a prison sentence or serious fines if proven. An applicant has the right of appeal to a Magistrates' Court if an application for a licence is refused and the government has promised to publish how this system will operate in future.

The power of the police will be extended under the new Act to include temporary events; the police will be able to close any premises or event "where disorder or noise nuisance is occurring". These powers can be both 'anticipatory or reactive'.

One other impact of the Act will be to end the 'two in a bar rule' which allowed two artists (a duo) to perform without a licence. The rule was not ideal but it was thought to be better than a 'none in a bar rule'. But incidental or background music (such as a pianist in a restaurant or elevator music) will not require a licence nor will certain other activities including:

- Morris dancing
- Use of a television or radio
- Entertainment in or from moving vehicles
- Entertainment in educational establishments for the establishment's educational purposes.
- Religious services
- Any of the licensable activities in a place of worship. (Note: a concert of religious music in a secular venue will still be licensable).
- Garden fêtes or similar community style events where there is no personal gain or profit.
- Films for promotional, education and information purposes.
- Film exhibitions in galleries and museums where it forms part or all of an exhibit.

The definition of incidental music is not clearly defined by the Act and is likely to be the subject of future controversy.

The view of the industry towards the new Act is that it is an attempt to cut down on red tape, inconsistency and over-charging. Huge amounts of attention and detail have been devoted to the problems of alcohol, pubs and clubs to the possible detriment of other areas, including outdoor events.

The Act is a new and untested piece of legislation; it will be interesting to see how workable and how effective it will be.

In 2003 the Department of Culture, Media and Sport set up the Live Music Forum that includes representatives from the music industry, the Arts Council, grass roots music organisations, local authorities, small venues and government.

The aims of the Forum are:

- To take forward the Ministerial commitment to maximise the take-up of reforms in the Licensing Act 2003 relating to the performance of live music.
- To promote live music generally.
- To monitor and evaluate the impact of the Licensing Act 2003 on the performance of live music.

The effectiveness of this new forum remains to be seen!

A problem facing our industry is that we are often assessed and regulated by authorities (usually local authorities) and enforcement officers who know nothing of what we do and who insist on applying regulations and conditions that are incorrect or inappropriate. In these circumstances the competence of such authorities and officers must be questioned!

4 THE HEALTH AND SAFETY AT WORK ETC. ACT 1974 (HASAWA)

Some of our Health & Safety law is very old and was put in place to stop us doing things like sticking small boys with brushes up chimneys (or something like that) and so that mill workers could come home from work with the same number of digits, arms and legs as they had when they left for work. Most of these laws have since been repealed but some of these old Acts are still in place!

UK Health & Safety law is now contained in various Acts of Parliament and also derives from case law and European Community directives. There is, therefore, no one single piece of legislation that can be regarded as *the* definitive statement of the law. It would not be appropriate to attempt to cover every aspect of Health & Safety law here and for this reason I have selected the main pieces of legislation and sets of Regulations which form the basis of the UK law and that are most relevant to our industry sector. The following chapters provide a fairly comprehensive overview of the current legal position but must not be considered the definitive legal position.

The HASAWA is an 'umbrella' act, which covers a great number of regulations, most of which apply to all the companies and services that make up our industry. The regulations covered by the Act include:

The Management of Health and Safety at Work Regulations 1999
The Provision and Use of Work Equipment Regulations 1998
The Manual Handling Operations Regulations 1992
The Workplace (Health, Safety & Welfare) Regulations 1992
The Personal Protective Equipment at Work Regulations 1992
The Health and Safety (Display Screen Equipment) Regulations 1992
The Control of Substances Hazardous to Health Regulations 2002 (COSHH)
The Reporting of Injuries, Diseases and Dangerous Occurrences Regulations 1995 (RIDDOR '95)
The Health and Safety (Information for Employees) Regulations 1989
The Employers' Liability (Compulsory Insurance) Act 1998
The Health and Safety (First Aid) Regulations 1981

The Noise at Work Regulations 1989
The Working Time Regulations 1998
The Electricity at Work Regulations 1989
The Lifting Operations and Lifting Equipment Regulations 1998
The Health and Safety (Consultation with Employees) Regulations 1996
The Health and Safety (Signs and Signals) Regulations 1996
The Work at Height Regulations 2004

The Health & Safety Executive can provide you with leaflets, books (both free and priced), and other easy-to-follow guidance and ACoPs on all of the above regulations and related subjects. Companies and businesses are advised to contact HSE Books for details of the literature available and to obtain copies of the relevant publications.

Ignorance of the law and the regulations is no excuse and won't get you off the hook if you should be unfortunate enough to run into trouble. The onus is on you to find out what your legal responsibilities are.

A list of useful publications is provided at the back of this book. They give more detailed information and you will certainly need some of these publications to enable you to carry out any risk assessments. This book is only a simple guide and only touches on the subject of Health & Safety at events. Having said that, I must admit openly to cribbing huge sections (it not the majority) of this book from various HSE and other Health & Safety publications (as I see no point in trying to reinvent the wheel), so it should be accurate.

The HSW Act 1974 forms the basis of UK Health & Safety law and its provisions should be considered before looking at any other relevant legislation. The scope of the Act is very wide and it applies generally to all workplaces, employers and employees, manufacturers, designers and importers of articles to be used at work, and even extends a degree of protection to the general public.

The Act is in four parts. Part I contains the most important provisions for people at work. Part II contains provisions concerned with the Employment Medical Advisory Service (EMAS), Part III has been repealed. Part IV contains miscellaneous provisions.

The Objectives of Part I of the Act are:

Securing the health, safety, and welfare of people at work.

Protecting people other than those at work against risks to their health and safety arising from work activities.

Controlling the keeping and use of explosive or highly flammable or otherwise dangerous substances, and generally preventing people from unlawfully having and using substances.

Controlling the release into the atmosphere of noxious or offensive substances from premises to be prescribed by regulations.

How The Objectives are to be Achieved:

1 Comprehensive Duties on Employers and Employees:

The Act requires "every employer to ensure as far as reasonably practicable, the health, safety and welfare of all his employees". The employer's duty extends to:

- Maintenance of a safe workplace, means of access and working environment.
- Provision of safe plant and safe systems of work.
- Preparation and review of a written statement on safety policy (very important).
- Identify and assess all significant risks (risk assessment has been a legal requirement since 1974, nothing new here!).
- Provision of safe storage and transport of substances.
- Ensuring that substances are able to be used and handled without risks to health and safety.
- Provision of training and supervision.
- Provision of suitable information and instruction.

2 Joint Regulation

Employees are actively encouraged to become involved in Health & Safety matters through recognised trade unions or simply as a matter of good employment practice. The Act established safety representatives and safety committees with statutory rights to be consulted and trained.

3 Employees' Duties

The Act placed statutory duties on employees to:

1. Take reasonable care for their own and other people's safety.
2. Cooperate with the employers to the extent necessary to allow those employers to comply with their statutory obligations, and
3. Not intentionally or recklessly interfere with or misuse "anything provided in the interests of health, safety or welfare in pursuance of any of the relevant statutory provisions".

4 Enforcement Agencies and Powers

The Act established a unified National Health & Safety Authority and a unified inspectorate. The Health & Safety Commission (HSC) and Health & Safety Executive (HSE) were created.

The Act provides for the HSC to have an element of self-regulation by giving it the power to approve codes of practice designed to give guidance on the practical application of the general duties in the Act. The HSC is responsible for new or updating existing laws, conducting research, providing information and dispensing advice.

The HSE, which is under the general control of the HSC, is the body principally responsible for enforcing the 1974 Act, on behalf of the HSC and under its guidance in relation to specified work activities. The HSE covers, amongst other areas, accidents that occur in factories, at docks, on the railways, or are related to industrial processes and agriculture.

The local authorities' jurisdiction covers such workplaces as shops, offices, warehouses, schools and colleges, hospitals and recreational facilities (including music and entertainment venues).

The Crown Prosecution Service (CPS) prosecutes manslaughter offences.

The 1974 Act empowers every enforcing authority to appoint inspectors whose powers are widely defined. In particular, they have the power to issue improvement notices, prohibition notices or physically seize and render harmless any article or substance that is a cause of imminent danger of serious personal injury.

Inspectors have the power to visit and enter a workplace or business premises without notice at any reasonable time. They have the right to talk to your employees and safety representatives and to take photographs, record findings and take samples or equipment. Inspectors can also inspect, copy and review documentation or take originals. They can direct that the premises be left undisturbed.

They can request interviews with anyone and ask them to sign a declaration as to the truth of the statement and they can do anything else they consider necessary for their investigation. If an inspector discovers a problem, he may issue either an improvement notice or a prohibition notice.

Improvement notices may be served on an employer or an employee. The notice states which statutory provision the inspector believes has been breached and the reason for this belief. It also states the deadline by which the breach must be remedied.

Prohibition notices may be issued where the inspector believes that a work activity involves or will involve a risk of personal injury. The notice prohibits the specified activities from being carried out until and unless the specified corrective measures have been implemented. A prohibition notice can take effect immediately or can be deferred to allow the recipient time to remedy the situation.

Health & Safety Executive inspectors can conduct interviews under the Police and Criminal Evidence Act but they do not have the power of arrest and cannot compel anyone to attend an interview under caution. If you are invited to attend an interview remember to:

- Give careful consideration as to whether to attend the interview or not.
- Consider the possibility of adverse inference being drawn if someone attends but does not answer questions put to them.
- No adverse inference can be drawn from a refusal to attend.
- Consider carefully who represents the company at an interview. For example, who can deal most appropriately with technical or financial matters aired.
- Interviewees can have someone present and this can be their solicitor.
- Remember the need to appear to be cooperative with the HSE at all times.
- Request advance disclosure of examples of questions to help preparation, ensure the most appropriate person attends and ensure prior collection of all relevant documentation.
- Section 20 of the HASAWA gives the authority to the HSE to ask to interview key witnesses following an incident. Key points to remember are:
    - Section 20 interviews can be verbal or you can submit a written witness statement.
    - You will be asked to sign a declaration confirming the answers you have given are true; you do not need to sign the declaration at the time of the interview.
    - It is often better to confirm that although you will not attend an interview you wish to be cooperative and submit a written statement in response to any questions.
    - Responses given to Section 20 interviews cannot be used against the person who gave them, thus protecting against self-

incrimination. If the HSE wishes to rely on the content of a Section 20 interview at trial, the witness needs to be called to give evidence under oath.

5. Health & Safety Regulations, Approved Codes of Practice and Guidance.

The 1974 Act laid the basis for the progressive replacement of all the existing legislation by modern Health & Safety Regulations and this is the primary means by which the objectives of the 1974 Act are to be achieved.

The Act provides a statutory framework for the implementation of European Community directives and the drawing up and enforcement of new regulations without any recourse to Parliament. Thus it is known as an 'enabling Act'.

Regulations are intended to set specific targets or define particular processes, which are required in law to be followed so as to achieve compliance with the main Act. These regulations can operate in two distinct ways:

a) They can require employers to manage matters in a particular way. For example, the Construction (Design And Management) Regulations 1994 lays down rules for the management of building works.
b) They can set specific, practical standards. For example the Health & Safety (Signs and Signals) Regulations 1996 state exactly which type of sign is suitable for fire exit routes.

To assist with compliance, Approved Codes of Practice (ACoPs) have been produced by the HSC for some regulations and the HSE has produced guidance on many of the regulations.

Approved Codes of Practice (ACoP) spell out how employers can comply with the Act and Regulations. It is not compulsory to follow an ACoP, but if you wish to manage a certain matter in a different way the onus is on you to prove that what you are doing is at least as effective in protecting Health & Safety. In reality you will be required to prove your systems are in fact more effective then those set in the AcoP, and this is probably almost impossible to achieve.

Guidance provides additional useful information to assist in complying with the law. As guidance is not written in the legalistic way that Regulations and ACoPs must be, the language is more assessable and the advice easier to understand and follow, therefore guidance carries less weight in law then ACoPs.

It is important to remember that the various regulations should not be taken in isolation from each other: many overlap.

The European Revolution

The original aim of the 1974 Act was for the repeal and replacement of all of the old legislation within two decades. However, by the mid-1980s most of the original legislation was still in force. In 1986 all members of the European Community signed the Single European Act. This gave renewed vigour to the 1974 Act's plans to replace the existing legislation, as it made it easier for the Community to introduce minimum standards for the health and safety of workers.

The UK has sought to implement EC directives (a directive is primarily intended to create legal relationships between the EC and the member state to which it is addressed) by means of statutory regulations made under section 15 of the 1974 Act.

In 1992 six new sets of EC Regulations were introduced which have had a significant impact on UK Health & Safety law. These main features of these regulations and the duties they impose are set out in a set of six guidance books and Approved Codes of Practice known as 'The Six Pack'. They are:

The Management of Health and Safety Regulations 1992 (revised 1999)
The Workplace Health, Safety and Welfare Regulation 1992
The Provision and Use of Work Equipment 1992 (revised 1998)
The Personal Protective Equipment at Work Regulations 1992
The Display Screen Equipment Regulations 1992
The Manual Handling Operations Regulations 1992

To sum up

The purpose of the Act is to improve health and safety at work. Specifically, the Act aims to:

- Secure the health, safety and welfare of persons at work including employers, employees, (whether full or part-time, temporary or trainee). It also covers the self-employed.
- Protect the general public against risk to their health and safety resulting from the activities of persons at work.
- Control the possession, storage and use of dangerous substances, such as highly flammable substances, toxic chemicals and explosives. Controls can apply wherever dangerous substances are found.

- Control the discharge of certain substances into the air. This includes:

  I. danger to health
  II. danger to the environment
  III. nuisance

Examples include pollutants, gases, vapours, particles and noise. Additional regulations apply to certain substances.

Under the Act, employers must ensure the health, safety and welfare of their employees, so far as is 'reasonably practicable'. This means an employer must provide:

- An assessment of any risks to workers' health and safety. This allows for risks to be eliminated or avoided, and for preventative and protective measures to be identified.
- Safe equipment and working methods, including safe machinery, appliances and tools. An employer must also ensure their proper maintenance and provide a safe system of work.
- Safe arrangements for the use, handling, storage and transport of all articles and substances.
- Information, instruction, training and supervision to help ensure employees' health and safety at work.
- Safe entrances and exits to and from the workplace. These must be maintained properly.
- A safe working environment that has adequate facilities and arrangements for the employee's welfare at work.
- A written Health & Safety policy including how the policy will be carried out, notice of any revisions and health surveillance for any risks that are identified.
- A safety committee, when required, and a competent Health & Safety assistant, advisor or department. Also, employers must consult with union-appointed safety representatives about the efficacy of Health & Safety measures.
- Protective clothing, equipment and safety devices, when necessary, at no charge to employees.
- Employers Liability Insurance to cover all employees.
- First Aiders and First Aid Equipment, an Accident Book and a means of investigating accidents and make reports where necessary to the HSE.

Employees also have duties, including:

- The duty to take reasonable care for the health and safety of themselves and others while at work. This means that employees are responsible for what they do *and* what they don't do!
- Cooperating with their employer to promote Health & Safety.
- Following Health & Safety policies provided by an employer and keeping up to date with any revisions, new requirement or regulations.
- Be aware of emergency procedures including fire evacuation, to follow them and also take part in fire and emergency training.
- To use any protective equipment and clothing properly, follow approved practices for cleaning and storing and report any damages.
- Practice safe work habits and obey safety rules.
- Report health and safety hazards promptly, including missing machine guards and safety devices, vandalised or missing signs, notices or instructions, blocked exits or escape routes and damaged or misused protective equipment and clothing.
- To report accident and near misses (I think that should be near hits rather than near misses!)

Duties of the Self-Employed

The self-employed are classed as employers under the Act because they employ themselves, they have the duties of both and employer and an employee.

Changes and Amendments

Towards the end of 2001 it was announced that the HASAWA was to be revised and that many of the regulations under the Act were also due to be updated, whilst updating the regulations is very common a complete revision of the Act is much more significant. I can only stress that it is really important to be working from the current edition of any regulations so it's vital to keep updating your information sources to keep abreast of any changes.

From time to time it is necessary to revise and update the various regulations under HASAWA, these changes may be major changes and revisions or very small, just a word or two has to be changed.

When very small amendments are made it is often not considered necessary to change the date of the regulations and produce new ACoPs or Guidance, instead several of these small changes are listed together as Amendment Regulations e.g. The Health and Safety (Amendment) Regulations 2001.

In most instances these minor changes are of no real significance unless a definitive legal viewpoint is required, then of course they are of very great significance.

5 THE MANAGEMENT OF HEALTH AND SAFETY AT WORK REGULATIONS 1999

This is perhaps the most important set of regulations; they are of very broad scope, covering all employers and all forms of work apart from sea transport. The centrepiece of the Regulations is the obligation on all employers and the self-employed to carry out a risk assessment under Regulation 3. The risk assessment must be 'suitable and sufficient' and must consider the risk to the health and safety of all employees and other persons arising from the conduct of the business. The regulations were again revised in 2003 to produce the Management of Health and Safety at Work and Fire Precautions (Amendment) Regulations 2003.

The purpose of risk assessment is to identify the measures which need to be taken in order to comply with the relevant statutory provisions. By complying with the relevant statutory provisions, risks will be reduced by default.

Many people make the fundamental mistake in thinking that risk assessment is all about the reduction of risk and make no reference to any regulations in their assessments. Don't make this common mistake!

The statutory provisions include the general duties under the 1974 Act as well as any more specific provisions.

Below is a summary of the responsibilities and obligations laid out in the Management Regulations.

Employers are required to:

- Carry out and record the findings of a risk assessment and any arrangements made as a result of the risk assessment (where the employer has five or more employees). [Regulation 3]
- Plan, organise, control, monitor and review all measures taken as a result of the risk assessment. [Regulation 5]
- Provide health surveillance. For example, keeping individual health records of employees. [Regulation 6]
- Appoint one or more 'competent persons' to help comply with the relevant statutory obligations. [Regulation 7]
- Establish appropriate procedures, which are to be followed in the

"event of serious and imminent danger to persons at work". [Regulation 8]

- Provide information on the risks employees are exposed to and the measures taken by the employer in accordance with the risk assessment procedure. [Regulation 10]
- Provide comprehensive information on health and safety to temporary workers and employees of an employment business. [Regulation 15]
- Consider the capabilities of their employees as regards health and safety before entrusting any tasks to them. [Regulation 13]
- Provide comprehensive information to contractors and self-employed persons who are working in a 'host' employer's undertaking on the risks to health and safety arising out of or in connection with the conduct of the first mentioned employer. [Regulation 12]
- Employees now have a duty to use any equipment provided to them by their employers in accordance with the instructions and training that has been given to them. [Regulation 14]

Regulation 12 has very serious implications upon event organisers and promoters as well as those who use the services of self-employed crew (contractors). You must provide them with copies of all the relevant risk assessments so they are aware of the hazards and the control systems that are in place that must be followed. This is where risk assessment becomes a two-way process, with information flowing in both directions.

With respect to **Regulation 7**, an employer may need to consider the use of a Health & Safety consultant to give the company the advice and information they need to help them comply, if they do not have a 'competent' person within the organisation. Specialist consultants are available who can provide such a service, but be warned: most safety consultants will prepare a Health & Safety policy and produce risk assessments, but most will not show how to set up Health & Safety management systems within your organisation, implement the policy, or show how to act on information provided with a risk assessment. If this is the case, find one who can advise you on such matters, take their advice, and act upon it. The fact you have a Health & Safety policy and risk assessments won't protect you when things go wrong: you can't be prosecuted for not having such items, but you can still be prosecuted for not managing safety.

A few years ago a company was fined £425,000 for a breach of Regulation

7 of the Management of Health and Safety at Work Regulations 1992 (as they were then).

The prosecution followed a fatal accident in which an employee was killed in the yard of one of the company's depots in an incident involving a forklift truck. In this instance the company in question was a supermarket chain, but proprietors and company directors in the production industry should take note: do you have 'competent' staff and 'competent' safety advisers or consultants?

The British Safety Council and NEBOSH Certificates and Diplomas are recognised Health & Safety qualifications of equal standing. It would be wrong to consider one above the other, and a competent safety adviser will hold a British Safety Council or NEBOSH qualification.

If a safety adviser or officer is appointed, it is vital that that person has sovereignty. They (and possibly other authorised persons) must be given full authority to instantly stop a show or halt work operations if safety is in any way compromised. In an emergency situation when lives are at risk, one person needs to make a rapid yet informed decision. Things need to happen quickly without long discussions and debates and that is not possible unless everyone understands and respects the safety adviser's position of sovereignty unequivocally.

Procedures for stopping a show should be worked out in advance and details must be supplied in writing to all persons involved in the 'show stop' procedure.

A safety adviser should have the following primary duties:

a) Monitor and review the implementation of the policy of the company or event, including inspections, checking the relevant certification, risk assessments and safety policies of contractors. It is not the role or responsibility of the safety officer or adviser to prepare policies, risk assessments or other documentation for contractors.
b) Provide advice and guidance on legislative requirements and safe working practices, to ensure safety procedures are followed and understood.
c) Make RIDDOR reports on behalf of a company or event.
d) Investigating accident occurrences, advising on the steps necessary to avoid recurrence.
e) Assist enforcement officers in their monitoring of safety standards within the company or event.
f) Liaison with Health & Safety Executive and local authority officers, and any other organisations with a view to improving any aspect of

health and safety.

g) The Health & Safety officer or adviser has the power to immediately stop any operations or practices which he considers unsafe, liasing with the appropriate persons as required.
h) The Health & Safety officer or adviser must be consulted in the planning and setting up of the event or new working practices to ensure safe systems of work are established at an early stage.

The Health & Safety officer or adviser must have no other responsibilities outside of his Health & Safety role.

The Health & Safety officer or adviser is not there to keep his client out of court or prison, *rather the Officer or Adviser is there to advise his clients about what to do to keep themselves out of court or prison*. You can take a horse to water but you can't make it drink! I have even heard of promoters telling safety officers to change reports to make the promoter look good!

Crowd safety officers or managers who deal with crowd management and dynamics need a separate and different qualification from an event safety officer or adviser. The subject of crowd management is not even touched upon within any of the existing Health & Safety qualifications and this situation needs to be addressed very soon. We must also consider production managers, for they are involved a great deal with health and safety and the contracting of 'competent' suppliers. Therefore, they also need a Health & Safety qualification.

Local authorities are now becoming much more concerned about the enforcement of all H&S regulations. They have a duty to enforce the HASAWA and the regulations it covers in addition to any entertainment licence conditions. This duty has been passed on to them by the Health & Safety Executive.

If they do not enforce the regulations, they may be considered to be negligent by a court of law if an accident has taken place. Therefore, they must enforce the law to cover their own backs. In fact, most Health & Safety work, as any Health & Safety practitioner will tell you, is just a back-covering exercise, but it's an exercise where you don't want to be left with anything left uncovered.

A Flow of Information

If genuine changes are to be made in the way we work, to bring us into line with current legislation, then we need to start with good systems of communication. There's no point in having great ideas without letting everyone else know.

Similarly, if you've got something important to say about the way you're working, there needs to be a good flow of information around your company so everyone can be informed of a risk or new work practice. (See Health & Safety, Consultation with Employees Regs.1996). There are no secrets in Health & Safety management and nothing commercially sensitive; every member of staff and others who may be affected must be given all the information. This is a legal requirement and even contractors must be informed of anything that may affect their health and safety, even if the problem is not caused or created by them but by others working alongside them.

A Healthy Attitude

We must aim for the establishment of what is known as a *Health & Safety culture* within the industry. That is to say, we all think about, consider and communicate with each other on health and safety matters at all times in a positive manner, that we try to raise awareness and change attitudes towards health and safety. At present we have a risk-taking culture.

To make this work and be effective, it must be done in a frank, positive and friendly way. Employers must involve their staff in H&S planning and management systems. To be effective it must be 'a way of life', a philosophy and part of the job. You must have the right attitude to health and safety as well as the paperwork.

When working as a production manager, tour manager or stage manager over the years I have had a few (one might even say ignorant) members of my crew say to me that I should leave Health & Safety stuff alone and get on with the job. Well, the answer to you is that this Health & Safety stuff *is* 75 - 80% of the job; it's not a bolt-on or added extra. I prefer not to work with such an attitude, but sadly we don't always have a choice.

What is a Safety Culture?

A safety culture is one in which safety is regarded by everyone as being an issue which concerns everyone.

As a result, safety rules are understood and adhered to, macho attitudes to safety *('hard hats are for wimps')* go out of the window and accidents and near-miss incidents are reported and investigated quickly and thoroughly. Strong safety cultures can be observed in many high-risk industries, such as offshore drilling installations.

The benefits of the safety culture to the organisation as a whole extend beyond a fall in the number of accidents. Studies have shown that companies demonstrating strong safety cultures also show improvements in performance, quality standards and staff morale.

Another example of this *H&S culture* in action (supplied to me by crowd safety manager, Jon Corbishly) is shown when safety professionals (such as Environmental Health Officers) arrive for an event-planning meeting on an empty 'green field' site that does not have so much as an overhanging tree branch as a potential hazard. The first thing these chaps do as they get out of their cars is put on hard hats and high visibility jackets. This is the 'norm' for them, their safety culture in action and, like it or not, sooner or later it will have to be part of our culture as well.

Developing and Promoting a Safety Culture

The approach your company takes to health and safety in the workplace will have a significant impact on the success of your policy. Many companies take a purely legalistic approach to the subject; they assume that legal minimum standards are acceptable and do nothing over and above what the law requires. This approach deserves criticism in that, through its own indifference to safety issues, this type of company is indirectly encouraging its employees to disrespect safety rules and safe working practices.

Companies which set the standards in Health & Safety management can demonstrate the existence of a safety 'culture' in their workplaces. This starts at base, in the office, warehouse or workshops. In many instances the 'core business' activities are out on site away from the office or warehouse, so it is often considered that the office and warehouse are drains on resources, but for health and safety purposes your 'culture' must start here: in the office and warehouse. Cultural factors influence all aspects of the running of an organisation including health and safety. Every organisation has its own unique culture or set of cultures, which develop and grow over time.

Through effective safety management it is possible to encourage and promote the development and growth of a positive safety culture in your workplace.

Like it or not, people are like sheep. New employees joining a business with no safety culture will adopt the attitude that Health & Safety is unimportant or to be avoided. Those who join a business with a strong safety culture will quickly follow and Health & Safety will be very high on their agendas.

It is generally accepted that the successful development of a safety culture in any organisation is dependent on the following:

- Acceptance of responsibility for Health & Safety from senior

managementand a clear commitment on the part of management to protecting the safety of all employees. This is best demonstrated via a company Health & Safety policy statement.
- Ongoing monitoring of safety in the workplace through hazard identification and risk assessment and the setting of realistic and achievable safety objectives.
- Thorough and systematic investigation of all incidents.
- Immediate action to remedy any deficiencies once discovered.
- The existence of a rewards system for good performance.

Rewards Systems

A major obstacle encountered by many safety managers is the apparent reluctance on the part of employees to respect and adhere to safety rules and safe working practices. There are three main reasons for this:

1. Workers who have been doing their jobs for many years without incident feel that they know best and management is interfering without really understanding the job involved.
2. The introduction of new rules and safe working practices may be perceived as an imposition by management, further restricting the autonomy of the worker. This may lead to resentment.
3. A reward scheme can help remove these obstacles and assist in the promotion of a safety culture in the workplace. Reward schemes may take various forms.

Safety Bonuses

The most visible means of rewarding good safety performance is financial. Offering a cash bonus to staff who show a clear and determined effort to raising safety standards in the organisation is a very effective method of encouraging safe working practices.

Competition

In larger companies, departments can compete against each other for a monthly or quarterly award – a trophy or flag, etc. Awards may be made for best performance, most improved performance, and so on. The winning department should be publicly praised through an award ceremony, for example, with members of senior management present.

Rewards

It is essential that effort and achievement are recognised and rewarded. Some form of reward such as a staff party an outing, or prizes should be offered by management for good safety performance and achievement of objectives.

Reward schemes such as these act to:

- Raise the profile of Health and Safety as an issue generally
- Generate interest and excitement in safety at work
- Raise staff morale through competition.

Planning for Safety

Health and safety starts at the early planning stage for any concert, festival, event or tour. Production managers/tour managers have a critical role to play throughout.

It is normally at this time that a production schedule is drawn up and this needs to be done with H&S in mind – to establish what are known as 'Safe Systems of Work'.

For example, a 'load out' after a show can be planned so that no one has to work on stage or can gain access to the stage when riggers are working above. Another example is allowing ample time for operations to be completed safely with time allocated for crews to rest, eat and sleep. On tours, production schedules should never be so tight that crews are expected to travel overnight between shows unless they travel in a properly equipped 'sleeper bus'. To achieve this successfully, staff, suppliers and contractors must be involved in this H&S planning. Our production schedule becomes a Method Statement for the event or show. As we will see later, a Method Statement is one of the 'controls' we can implement to reduce the risks associated with the possible hazards we may encounter in our work activities.

Most of the dangerous operations we deal with are during the build-up and breakdown of an event or show. Adequate health, safety and welfare facilities must be put in place ready for when the first crew arrives and must remain in place until the last of them has left.

I'm sure we all know of the well known festival promoter who turns off the water supply on his site and shuts down the first aid posts only a matter of hours after the last band has finished playing and when hundreds or even thousands of crew are still packing and cleaning up on site for days after. They are left with no water, no toilets, no washing facilities and no first-aid cover, a direct breach of the Workplace Health, Safety and Welfare Regulations

1992, the Health and Safety (First Aid) Regulations 1981 as well as a breach of the HASAWA itself.

Responsibilities

It will become more and more obvious that we *all* have a large number of responsibilities under the HASAWA.

We start with the venue or site owners who have a responsibility to make certain that their site or venue is safe and free of hazards for anyone they invite into or onto their premises. In particular, they must insure adequate means of access and egress and also insure that any item of plant or any chemicals or substances are safe and without risks to health of the persons working there.

They must inform the person that hires the site or venue of any known hazards, they *may* also have an implied duty to ensure that their client is competent and operating in a safe way according to the HASAWA.

Next in line are the promoters or acts/artist management and service companies. They also have a legal duty under the HASAWA to provide a safe working environment, to ensure they are 'competent' and can prove they are 'competent' and to ensure that the contractors they appoint are 'competent' and have suitable and adequate H&S management systems in place.

The latter can be done by asking for a copy of a company's Health & Safety Policy Statement, copies of any written risk assessments, method statements, records of training, qualifications, accident records etc. which can then be checked to see if the proposed controls are considered to be suitable and adequate for the proposed job. This duty will almost certainly fall back on to production/tour managers, as it is normally they who book suppliers and contractors, but somebody also has to book them and check their Health & Safety competence.

It is advised that at the same time they request a copy of the contractor's public liability/employers' liability insurance to see that that is also adequate and to follow up references. Records of this assembled information should be kept on file.

Employers, promoters and venue owners should consult the Guidance on Regulations for The Provision and Use Of Work Equipment Regulations 1998 and The Approved Codes Of Practice for The Work Place (Health, Safety and Welfare Regulations 1992 and The Management Of Health and Safety at Work Regulations 1999). These give full details of some of the responsibilities these groups of people have and must be considered essential reading.

An employer has the greatest responsibilities to comply with all the relevant regulations. I would advise using the services of a Health & Safety consultant to carry out an audit, the results of which can be analysed to produce a prioritised plan of action that can be implemented and auctioned over a scheduled period of time; it will probably be impossible to do all that is required to meet the regulations all at once.

The HSE realises that not everything can be done at once and will do anything possible to encourage and assist a business that is trying to get it right. However, they will not tolerate a company that takes no action and who waves two fingers at Health & Safety legislation.

Finally, as mentioned, it is a legal requirement under the HASAWA 1974 and the Management of Health and Safety At Work Regulations 1999, to produce a written Health & Safety Policy Statement and Risk Assessments if you employ five or more people. However, under the Management of Health and Safety At Work Regulations 1999, if a significant risk could be created to a large number of people, a risk assessment, written proof of a safe system of work and training would still be required from a business with fewer than five employees. Self-employed truss monkeys, take special note!

While not mentioned in any regulations, employees and self employed persons have a responsibility to be fully prepared and fit to start work, they should ensure that they are fully rested before work starts and not still drunk or hung-over, and they should have eaten a sensible meal and be suitable dressed to undertake they type of work they are required to do.

Jewellery should not be worn and long hair must be tied back if it is likely to be a safety hazard. Laminated passes and wristband passes can also be safety hazard in some situations. Employers must ensure that regular rest and meal breaks are planned in the schedule, and that adequate toilet, washing, rest, first aid and welfare facilities are available.

These days, artists and productions travel internationally, and the information contained in this book is based upon UK Health & Safety legislation that is UK specific, with a great deal of it originating in the UK. All foreign artists and productions should have their own Health & Safety standards, although it may not be as uniform or coherent as the HASAWA.

Promoters, production managers or those using the services of foreign artists or production should stipulate that at least one person from the artist's party or production can speak and read English, and ensure that this person passes on all the relevant safety information about the tour or show. In practice, standards may sometimes fall short of UK requirements and close monitoring of the work practices may be required. They must, however, conform to UK Health & Safety laws. Your Health & Safety adviser should consult with foreign artists and suppliers for the outset to ensure compliance.

6 ACCIDENT COSTS

Ultimately, it's the employer's responsibility to see that the law is being complied with (not forgetting that a self-employed person is also an employer because he employees himself), but everybody risks being prosecuted and, again, ignorance is no excuse.

When health and safety is neglected, it's going to hurt. As well as the injuries to people, it's also going to hurt someone's wallet. There can be huge hidden costs after a serious accident and these are usually ten times greater than the visible costs which may only be the tip of the iceberg. The hidden costs include lost fees, loss of productivity, increased insurance premiums, internal and HSE investigations, loss of reputation, 'down time' on hire equipment, and civil claims. It's better to spend a few poubds now getting Health & Safety management in order than to lose everything when things go wrong. It's a proven fact that one accident claim can cost far more than the cost of putting an efficient safety management system in place.

In 2002 the government announced new proposals to extend the current practice of recovering NHS costs from those responsible for causing road traffic accidents to include all personal injury claims. Under the new proposals the NHS would be able to recover the cost of hospital treatment from private companies and public bodies. As a result, individuals and organisations that pay compensation to employees for an accident relating to their work activities will also be liable to pay for any NHS treatment provided to the injured person. It is estimated that this will generate £120 million per year for thc NIIS as well as acting as a major incentive for employers to establish effective safety management systems and to comply with Health & Safety law.

In 2003 further announcements were made in a proposed Health & Safety (Offences) Bill that will seek to increase the maximum penalties that can be set by the courts. This would:

- Extend the statutory £20,000 maximum fine available in the lower courts to a much wider range of Health & Safety offences that currently only attract a maximum penalty of £5000.
- Make imprisonment more widely available for most Health & Safety offences, and

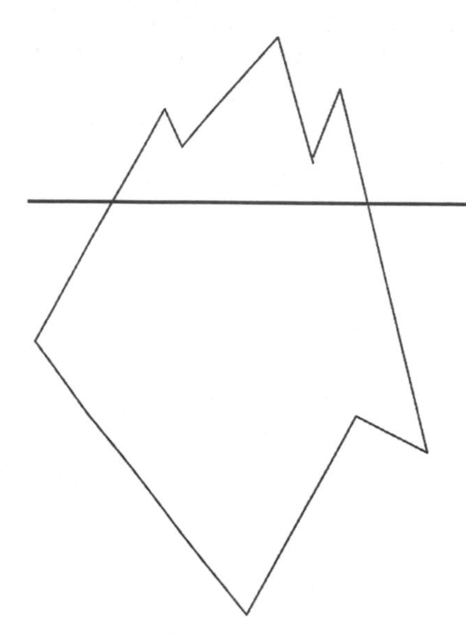

Tip of the Iceberg
The huge hidden costs of an accident are below the waterline.

VISIBLE COSTS
Injured person/s
Prosecution/Fines/Imprisonment

HIDDEN COSTS
Loss of reputation – bad image – bad PR.
Lost productivity and sales
Loss of staff morale
Increased insurance premiums
'Down time' on staff and hired equipment
Civil claims
Legal fees and costs
Lost time and costs during investigations
Medical costs
Compensation payments
Damaged plant and equipment
Overtime wages to make up lost time
Replacement labour – recruitment – training
Replacement equipment and parts.

- Increase the fine for not having a valid employers' liability insurance policy to £20,000.

One observation I have made over the last couple of years is that attitudes in our industry are changing towards Health and Safety. The old school of thought was 'how do we avoid or get away with this?' The new view is: 'what do we need to do to comply?'

The latter view is the only one we can adopt if we are to survive. My own small business would not have obtained work from local authorities and the National Trust if we did not comply.

It's going to take longer to take effect in the rock 'n' roll side of the business and some of the paymasters will almost certainly resist to the bitter end, but I don't think they will win. All we are trying to do is stop people getting hurt.

So, planning, procedures, hazard identification, risk assessment, exchanging information, record keeping, monitoring and review of your control systems is now the order of the day.

To be effective, Health and Safety, like all other parts of a well-run quality business, has to be correctly managed. In order to achieve this, it's going to mean extra work for production managers and employers.

If you want to be a quality business then you have to do it. Not only that: the law says you have to do it. If you don't agree, then don't blame me! I am only trying to explain the law – I don't make the law.

Showing the 'Yellow Card'

Managers and employers should not hesitate to issue warnings to contractors or employees if unsafe work practices are seen to be taking place.

The severity of the warning should correspond with the level of risk (as shown in the risk assessment for the operation).

For using unsafe working practices in a situation that could be potentially life- threatening, the manager or employer should have the work stopped until the situation has been rectified.

A lesser breach will require a formal letter of complaint to the company/ employee or freelancer in question and a minor breach may only require a few words with the crew involved. Despite our industry's reputation with drink and drugs, the use of alcohol and recreational drugs by crew members is a risk to safety and is, therefore, an instantly sackable offence.

As with everything else to do with Health and Safety, details of all complaints and reprimands must be recorded.

Should you find yourself being prosecuted by the HSE or the local authority for a breach of the HASAWA, you can console yourself by remembering that at present a £20,000 fine and/or six months in prison may be imposed by a Magistrates' Court; a Crown Court may impose an unlimited fine and/or two years in prison for breaking Health and Safety regulations; and these penalties are due to be increased very soon. After these possible penalties there may be civil claims made against you, even before we have Corporate Killing on the statute books. This is a new offence intended to make companies accountable in criminal law when they fall far below the standards that could reasonably be expected in the circumstances. The proposed maximum penalty would be an unlimited fine and an order to correct the original cause of any accident.

Further to corporate killing, there are another three new offences. Firstly, Reckless Killing: where a person (including a company representative) was aware of the risk that their conduct would cause death or serious injury, knew that in that situation it was unreasonable to take that risk, but did so anyway. Maximum penalty: life imprisonment.

Next we have Killing by Gross Carelessness: where a person (including a company representative) should have been aware of the risk of causing death or serious injury. Maximum penalty: ten years' imprisonment.

Killing when the intention of a person was to cause only minor injury but death was caused by an unforeseeable event. Maximum penalty: between five and ten years' imprisonment.

And if that's not enough, then after prosecution you will find full details of your offence, including your name and address, published on the HSE website at: www.hse.databases.co.uk

This database already lists several companies from our industry which have been convicted of Health & Safety crimes, so don't get yourself listed.

Most people assume it will be the HSE that will bring about any prosecution for breaches of Health & Safety law, but in fact it will be the local authority who have the duty to enforce Health and Safety regulations in any theatre, club, festival site, sports hall or other venue used for entertainment.

It is also the case that the enforcing authority will rarely prosecute for a breach under the regulations such as LOLER or PUWER. It is much more likely they will use the HASAWA as it's much easier for them to enforce and bring about a successful prosecution.

The Other Side of the Coin

Of course it's not always the employee or self-employed contractor who is in the wrong; unfortunately it's often the managers and employers who have got it badly wrong or just don't know better and who give instructions and orders to staff and crew that *do* know better.

It is they who find themselves in a real dilemma as they don't want or can't afford to upset the employer or client by refusing to carry out the order or instruction, even when they know it's unsafe, breaches safety regulations and puts people at risk.

It seems almost daily now that I hear of problems from self-employed persons and employees who are required to do things that are blatantly unsafe yet, when they challenge the manager or employer giving the order or instruction, they are told to "get on with it or go get another job". They dare not risk losing the job as, not only may other work be in short supply, but their reputation is also on the line and they don't want to gain a reputation for being 'difficult and obstructive'.

The excuses I have heard by employers and managers for ignoring Health & Safety are getting incredible and I list a few examples here. Not only that, but I know of one tour manager who has taken his client to court for a breach of contract: his client dismissed him because he said it was unsafe for the crew to drive themselves between shows overnight with no opportunity to get any sleep!

Examples:

"Health & Safety laws do not apply to our industry."

"No-one joined our industry to get Health & Safety."

"When you work for us you do as you are told."

"How on earth am I going to crew my shows if I start insisting that self-employed workers have their own insurance?"

"Our insurance company says it's not true."

"It's all very well on paper, but we simply don't have the time to do it all that way."

"We can't go bothering the client with all that – they are spending a lot of money here."

"Even if I don't have my own insurance, I'm still covered by my employers."

"Don't worry, our clients are in the audience, we are not going to do anything to harm them."

My personal favourite is the well-known festival promoter who said: "We don't bother too much about Health & Safety here, it all seems to work and our crews have the Dunkirk sprit." What a disgusting comment! Forgive me if I am wrong, but I thought we were at work and that the war was over. Expecting the crew to work hard is fair enough, but this promoter also expects his crew to live in a tent and work a twenty-four hour day in mud up to their necks and to have to eat food from catering that is only just passable for pigs to eat. This is exactly why we have Health & Safety regulations! The same festival promoter also said that the only health and safety problems he encountered were all the Health & Safety rules.

The self-employed sometimes have themselves to blame and put themselves at risk when they take on too much work because: "It's a regular client and I can't let them down" or "It's a new client and they may put a lot of work my way" or "It's not a regular client, but they do pay well and on time." Remember, even the self-employed have to conduct risk assessments on working hours and stress even though the Working Time Regulations do not apply to them. There has to be a time to say "No", and this should be before you fall asleep at the wheel of your car on the way home from a gig.

So what do you do when you're working on a show and you are told to do something that you know is dangerous and putting people at risk? Well, the first thing to try and do is to have a calm, logical and sensible conversation with the person giving the order or instruction and attempt to show them clearly the dangers and risks involved. Hopefully this will do the trick.

What if that does not work? Well, if you're an employee perhaps you can call your company's office and tell them what has taken place. An example of this was supplied by lighting designer, Dave Smith.

An employee on a rock tour in Spain was told by the production manager to build the ground support system with one leg on a stack of beer crates. The crew member refused and was promptly told he would be sent home and that his company would be informed that he refused to carry out the instruction.

The employee in question went to a phone, called his company in the UK and explained what had taken place. The company immediately agreed that what had been asked was dangerous and sent a fax to the production manager informing him that their employee acted with the full support of the company and that their employee was not to erect a ground support on beer crates. At this point the production manager backed down and an alternative and safe solution was found. A good result.

The other option, and one that self-employed contractors will probably have to use, is to draw up a risk assessment and present this to the production manager or whoever asked for the task to be carried out. With the risk assessment should be a statement for the person giving the instruction or order to sign stating that he or she will take full responsibility should anything go wrong. It has been proven that not many people are prepared to sign such a statement and that an alternative system is usually found very quickly.

The last two options are the extremes. We have a legal duty not to put people at risk and we should refuse to do the job which will almost certainly mean getting the sack, but you will have a good case for unfair dismissal, especially if you can get any evidence in writing, you have a witness and you have kept a few notes of what took place and what was said and by whom; you can also report the incident to the HSE, who will treat it all in confidence. The HSE have a 'whistle-blowers" charter for just these kinds of situations.

Whatever route is taken, try to remain calm and logical; don't enter into arguments. Saying "no" is always difficult and no one wants to lose work or reputation, but a sensible solution can normally be found with a little thought.

Keeping Records

I make no apologies for stressing the need to keep records for everything to do with Health and Safety, regardless of the size of your business or organisation. This includes:

- Personnel Files with details of staff selection procedures, CVs, references, interviews, training records, information and instruction given.
- Equipment test and inspection records (including operator checks)
- Maintenance and safety inspection records
- Details of certificates of competence and qualifications
- Risk Assessments
- Method Statements
- Permits to Work
- Warnings and Reprimands
- Insurance
- Health & Safety Policies
- Accident/Dangerous Occurrence records and reports
- Official Procedures
- First Aid equipment and arrangements
- Site/Venue Checks
- Fire, Evacuation and Emergency Procedures and Equipment
- Employee Consultation Records
- PPE Checks

These records form a major part of your quality assurance control, it demonstrates your quality management system and procedures. Records can help to prove competence and will be required as evidence in a court if things go wrong.

In some cases it is a legal requirement to keep certain records, and it is equally important for the self-employed. A study of your records and control systems will enable an H&S audit to be carried out on your business.

It should be common practice to ask for risk assessments, accident records, policies, method statements, insurance details etc. from a company that is *tendering* for a job.

This information obtained from assessments etc. enables an employer to check the proposed controls; they then can be rechecked during an operation to see how they are working. Requesting, compiling and checking this information is a good H&S management system and is an example of what is required of production managers and employers.

An accident is an unplanned, uncontrolled event leading to injury, property damage or a near hit.

7 BUT WHAT IS A 'SELF-EMPLOYED' PERSON?

The HSE considers that a self-employed person is someone who works for him/herself and that a self-employed person may also be an employer. The fact that an individual pays his/her own income tax and national insurance does not necessarily make them self-employed.

HM tax inspector has a list of indicators towards employment or self-employment. The more indicators towards self-employment, the more likely they will consider you to be self-employed.

If you work for an employer who covers you under his/her employers' liability insurance, then the chances are you are an employee even if you pay your own tax and national insurance contribution. This is even more the case if you work under supervision, such as a 'stagehand/local crew' person may do.

Employers need to take care as well – you may end up with a large bill for income tax or national insurance contributions for your so-called 'self-employed' crew who you thought were paying their own tax and N.I. contributions. This has already happened to at least two companies in our industry and it cost one of them a fortune and made the other bankrupt. It is for this reason that some companies insist that all staff are on PAYE even if they only supply services on the odd day or occasion; other companies will only engage the services of self-employed persons who have become Limited Companies. However, this may make the Limited Company susceptible to the IR35 Tax Regulations in many instances, and so some accountants now advise clients not to become a Limited Company unless they are making a profit of over £100,000 per year that will cover the increased costs and responsibilities. How many self-employed crew make that kind of money? And with every Budget, the regulations get tougher as the 'loopholes' get closed.

Employers/crew bosses, take note: you may have employees you did not think you had, together with all the Health & Safety, tax and national insurance responsibilities for them! The more you provide for contractors, the more likely they may be classed as your employees.

Take for instance the 'self-employed' sound engineer who goes on tour with a band for a few months – he is paid a good wage, provided with transport, accommodation and food and is probably covered by the tour's public/ employers' liability insurance. The HSE and the Inland Revenue say this engineer is not 'self-employed' but an employee (and therefore should be 'on the books' and paying tax by the PAYE system).

Self-employed persons also take note: are you *really* 'self-employed'? Who provides your public liability insurance cover?

If you don't have your own then you are probably an employee – you had better check your employers have the necessary insurance cover for you.

Obviously, it is very important to ascertain an individual's position as to whether they are in fact employed or self-employed, as the law sees it.

No employer in his/her right mind should engage the services of self-employed persons unless they carry public liability insurance suitable for the operations they intend to carry out, unless the employer intends to cover these individuals by his/her own employers'/public liability insurance. In which case, they may then become his/her employees and the employer has to take all the responsibilities of an employer under the HASAWA and other employment laws.

There are some other factors which may be used to decide if you are an employee or not and the Inland Revenue uses these simple checks (contained within the IR56 Tax Regulations) which can be applied that will give indications:

Indications towards being employed:

- Your employer deducts income tax and national insurance from you at source.
- You work for long periods for the same client.
- Your client supplies all the tools (other than just small hand tools) and materials.
- Your client provides employers' liability insurance cover for you.
- You work under supervision.
- You can't sub-contract or assign the job.
- You are told where to work, how to work and what time to start and finish work.
- You have a 'contract of service'.
- You expect your client to give you work and he expects you to accept the work.
- You work for a weekly, daily or hourly rate.
- You don't risk your own money and there is no possibility that you

will make a financial loss.
- You work at the premises of the person you work for, or at a place or places he or she decides.

Indications towards being self-employed:
- You don't work under supervision.
- You work for a number of different clients.
- You provide your own tools (other than just small hand tools) and materials.
- You have a 'contract for services'.
- You hold your own public liability insurance (and employers' liability insurance if you also employ people).
- You can sub-contract the job or provide a deputy to carry it out.
- Your contract does not have 'mutuality of obligation', i.e. on an employer to provide work, and a worker to accept it.
- You employ other people and pay them out of your own pocket.
- Within the overall timescale of the project, you can choose how and when the work will be carried out.
- You have the final say in how the business is run.
- You have to correct unsatisfactory work in your own time and at your own expense.
- You stand to profit from the good management of the business.
- You use 'entrepreneurial skills', in other words, you take business risks.

It is possible for you to lose money as well as make it. You are paid an agreed amount for the job regardless of how long it takes you.

One of the checks that employers can make to ascertain that the 'labour only contractors' they intend to appoint are genuine self-employed persons, is by getting them to quote for the jobs available and not offering a monthly, weekly, daily or hourly rate.

Quoting for a job uses 'entrepreneurial skills' and is an 'acid test' for true self- employment, but even then it may be possible to do the occasional job for a daily or hourly rate. For instance, if it proves to be extremely difficult to quote a price to do the whole job, because until you start you don't know how long it's going to take or what is involved. This seems to be acceptable as long as it's only occasional and does not represent the majority of your work.

It is a trend within the film industry and in theatre-land to put freelancers 'on the books' for jobs lasting more than a week or so, and deduct tax by the PAYE system; this is perhaps the only way to deal with long-term freelancers who are known as PAYE Freelancers. However, I personally feel the correct term is 'short-term employees'. I suspect we are going to see more of this in our industry as many so-called 'self-employed' on a long job or tour really should be PAYE employees. Some freelancers who have numerous clients also opt for PAYE status with each client. This is quite legal and acceptable, in fact you can go from PAYE employee to self-employed and back again from job to job; employee in the morning, self-employed in the afternoon.

I should point out here that 'freelance' does not mean the same as 'self-employed', as 'freelancers' can be PAYE employees of a number of clients.

The only body which can really decide if you are genuinely self-employed or not is the Inland Revenue, and even then it's quite possible for each local office to give a different definition or ruling on each individual case. All the information I have given here acts only as an indicator, but it's a fair bet that the more indicators you have pointing towards self-employment, the more likely you are to be self-employed. I would advise employers to take steps to ensure that any self-employed crew they use are acceptable as self-employed to the Inland Revenue. Your local tax office will advise you.

Employers must remember that genuine self-employed persons are not 'short-term employees' and must not be treated as such.

Employment status is not a matter of choice. People are self-employed if they are in business on their own account and bear the responsibility for its success or failure.

The Inland Revenue quotes that: "The tax rules for self-employed people are designed to reflect the day-to-day transactions of the true risk-taking entrepreneur. Genuine self-employment is about being in business on one's own account and not simply applying a label of self-employment".

Bear in mind that because you are self-employed in one job doesn't necessarily mean you will be in your next job.

The HSE in a statement contained in the *Event Safety Guide* on employees and self-employed persons for the purposes of health and safety is as follows:

If people working under the control and direction of others are treated as self-employed for Tax and National Insurance purposes they may nevertheless be treated as employees for health and safety purposes. It may therefore be necessary to take appropriate action to protect them. If any doubt exits about who is responsible for the health and safety of a worker this could be clarified and included in the terms of a contract. However, remember a legal duty under Section 3 of the Health and Safety at Work Etc. Act 1974 cannot be passed on by means of a contract and there will be duties towards others under Section 3 of the Health and Safety at Work Etc. Act 1974. If such workers are employed on the basis that they are responsible for their own health and safety, legal advice should be sought before doing so.

Confused?

8 THE WORKPLACE HEALTH, SAFETY AND WELFARE REGULATIONS 1992

These Regulations apply to employers, occupiers of factories, and persons in control of workplaces in connection with the carrying on of a trade, business or other undertaking, whether for profit or not. A concert venue or a festival site is certainly classed as a Workplace under the Regulations, as are offices, workshops and warehouses.

The Regulations require:

- Workplaces, equipment, devices and systems to be maintained in an efficient state, in efficient working order and in good repair [Regulation 5].
- Effective ventilation for enclosed workplaces, with visible or audible warning of problems [Regulation 6].
- Temperature control and the provision of thermometers in all workpaces [Regulation 7].
- Suitable and sufficient lighting in all workplaces (Regulation 8).
- Cleanliness of all workplaces and the removal of waste materials [Regulation 9].
- Adequate floor area, height and space in workplaces [Regulation 10];
- The provision of suitable workstations both for the person at work and for the work that is likely to be done there [Regulation 11].
- Safe floors on 'traffic routes' – 'traffic routes' includes any route for pedestrian traffic, vehicles, or both, stairs, staircases, fixed ladders, doorways, gateways, loading bays or ramps – [Regulation 12].
- The provision of fences, covers and other safety precautions to prevent persons or objects falling in the workplace [Regulation 13].
- The provision of safety material or protection against breakage of windows and other transparent surfaces in the workplace [Regulation 14].
- Safe and adequate traffic routes in the workplace, for example, adequate turning space for vehicles and the separation of vehicles from pedestrians [Regulation 17].

- The safe construction of doors and gates [Regulation 18].
- Escalators and moving walkways to function safely and to be equipped with safety devices and one or more emergency stop controls [Regulation 19].
- Rest rooms and rest areas to have facilities, which protects non-smokers from discomfort caused by tobacco smoke.

It is the employers' responsibility (under the Workplace (Health, Safety and Welfare) Regulations 1992) to ensure that the working environment is healthy and that suitable welfare facilities such as toilets [Regulation 20], washing facilities [Regulation 21], drinking water [Regulation 22], accommodation for clothing [Regulation 23], facilities for changing clothes [Regulation 24] and facilities for rest and to eat meals [Regulation 25] are provided for staff. These should be segregated into smoking and non-smoking areas or suitably ventilated to prevent the discomfort and hazard of passive smoking. Under the law, it now appears that non-smokers have rights; smokers on the other hand have no rights.

The Regulations also require that room temperature should be at least 16°C (after the first hour) where people work sitting down (e.g. offices) and at least 13°C for an indoor workplace where staff are engaged in active work. If it is not possible to maintain that temperature, then employers must provide body-warmers or other suitable clothing for workers.

A thermometer must be available to persons at work to enable temperatures to be measured throughout the workplace.

For further details see the Approved Code of Practice on the Workplace (Health, Safety and Welfare) Regulations 1992.

Areas of Responsibility for Venue Owners and Managers

Venues and venue managers also have responsibilities and an agreement should be made within the initial Tenancy Contract between the venue and promoter for exact areas of responsibility regarding provisions of service and facilities (i.e. cleaning, security, car parks, toilets, etc.). Some areas of responsibility may overlap and will vary between venues and between different shows or tours. These agreed areas of responsibility are for the specific show only and do not absolve any other statutory responsibility under law.

It is the collective responsibility of both venue and promoter to ensure the suitability of the venue for the proposed event. Depending on the nature of the event you may need more than the existing facilities than the venue has to offer, but this and the hazards associated with the event will become evident from undertaking a suitable and sufficient risk assessment.

The venue is responsible for providing a safe place of work and environment for its staff, contractors, visitors and clients - the promoter, so far as is reasonably practicable.

The venue is directly responsible for the upkeep of the venue *fabric* – i.e. roofs, walls, columns, floors, ducts, heating and ventilation systems, public gathering places and thoroughfares, basic fire-fighting cover, and all plant, materials and equipment normally contained therein which has not been expressly brought in or hired directly by a promoter for a specific show or event. If purpose-built, static restrooms, toilet, hygiene and welfare facilities for visitors are provided, then the venue is also responsible for their maintenance.

It is the venue's responsibility to satisfy itself, so far as is reasonably practicable, that suitable arrangements are in place and to ensure that the promoter has developed a full Health & Safety Management Programme, including a suitable and sufficient risk assessment for the show or event, in order to fulfil the venue's legal duties as set out in Regulations 3 and 4 of the Management of Health & Safety at Work Regulations 1999.

The venue must ensure that all its staff are trained and that the published emergency procedures are adequately communicated to all parties using their facilities onsite.

The venue should agree and implement suitable temporary management hierarchy so as to cover all the risks assessed and any emergency procedures required for each specific event, which may vary from event to event. A 'Health & Safety file' should be kept for each event, containing such documents as the risk assessments and statutory forms for RIDDOR, COSHH, etc.

The venue has a responsibility to pass on any information relating to any site-specific Health & Safety requirements/procedures to promoters, who then should disseminate this information down to the Production

Step-by-Step Safety Checklist for Venue Owners and Managers

It is your duty to promoters to ensure, so far as is reasonably practicable, that the venue is a safe one in which to hold the proposed event. The following checklist is suggested:

1. Appoint a 'competent person' or 'safety 'coordinator' for the planned event: This person will take responsibility for managing and coordinating Health and Safety matters both pre-show and on site between you and

the promoter, and you and the local authority. He/she can be an experienced and competent member of your own staff or from an independent body.

2. Review the accident statistics with the promoter: From the details of the promoter's last show/event/tour, if applicable, establish the most common type of accidents or incidents and enter these in your risk assessment for their forthcoming event. Note too if there were any serious accidents or near misses and bring these to the attention of the organiser. If the artist is new, reviewing the general accident records with the promoter should help to establish any patterns in accident types. If the act is an established one but just new to your venue, make available the existing accident records in order to establish the necessary control measures.
3. Undertake a separate in-house risk assessment: This may be done in conjunction with the local authority as part of an entertainment licence application. Consider the hazards at your venue which present risks to your in-house staff and contractors who work on your behalf within your venue. This risk assessment should also be instrumental in creating a safe working environment in which the promoter can hold their event.
4. Undertake a risk assessment for the show. From the accident records and details of the show supplied to you by the promoter, consider the most significant hazards and risks that will be present on-site throughout the three areas of the event, i.e., the 'get in' and build-up, the show and breakdown and 'get out' phases.
   Undertake this with a senior member of the promoter's or production team and agree who exactly will be responsible for on site, and the control measures in order to be implemented by each organisation.
5. Notify all parties concerned of your findings: From the risk assessment you will be aware of what the promoter expects and what you are to provide for the show (such as better hall lighting, external signage, traffic management, etc.). All your in-house staff and tied contractors must be made aware of the safety control measures you are implementing, usually by way of: a) safety bulletins; b) staff training; c) staff meetings; d) 'toolbox' talks; and e) safety notices or flyers distributed on notice boards, etc. The promoter should receive written notification of precise areas of responsibility and any guidelines on specific venue requirements.

6. Compile a ‘Health & Safety filing system. This should be kept up to date and contain the following for each event:
    - Tenancy Agreement
    - Risk assessment for the show
    - Accident records and reports
    - Policy statement
    - Pre-show timetable (showing dates of all the planning meetings such as production meetings, promoter meetings, etc.)
    - Example stage and rigging plans
    - Sample documents you produced for the promoter, i.e., safety information
    - Promoters material
    - Post-show report.
7. Chase and collate, from the promoter:
    - General risk assessment for the event if not done with you
    - Structure Drawings and Rigging Plots for each such stand or feature
    - Structural Calculations and Inspection Certificate for each such structure
    - Method Statement for each structure and rigging operations
    - Details of any item, material or chemical that has been brought to the promoter's attention by way of the production's risk assessment
    - Health & Safety Policy Statement
    - Copies of the production risk assessments as required
    - All other relevant information that you may stipulate in the Tenancy Agreement.
8. Review the event: A short while after the event, it is suggested that you evaluate everything that went right – and what went wrong – to enable you to plan and initiate control measures for the next show or event. Revise the existing control measures if necessary for the next show at your venue, and relay these findings to all relevant parties.

Learning from experience is not good enough - we must learn safety and not work by the old saying: "We have always done it that way."

9 THE PROVISION AND USE OF WORK EQUIPMENT REGULATIONS 1998

These Regulations impose many obligations and warrant careful study. Regulation 7, which requires the employer to restrict the use and maintenance of virtually all 'work equipment' (other than livestock, substances, structural items and private cars) to specific persons, is likely to have wide implications. In addition, Regulation 13, which increases the protection against burns, has universal application, and is a substantial increase in the protection afforded. The Lifting Operations and Lifting Equipment Regulations 1998 (LOLER) are closely linked to The Provision and Use of Work Equipment Regulations 1998 (PUWER). The Approved Codes of Practice for these two sets of regulations should be read in conjunction with each other.

The definition of work equipment is very wide and can include items as diverse as a pair of scissors, a ladder, a forklift truck and a rigging motor. (Work equipment is defined as any machinery, appliance, apparatus or tool and any assembly of components, which, in order to achieve a common end, are arranged and controlled so that they function as a whole).

The Regulations Require Employers:

To ensure that work equipment is safe and suitable for the purpose for which it is used or provided (has been risk-assessed) and to ensure that the equipment is only used for the operations for which it is suitable [Regulations 5 (1), (3)]. For example, a forklift truck should only be used for lifting and moving objects, it should not be used for towing other vehicles or items of plant.

To provide adequate Health & Safety information and training to those who use the equipment and to those who supervise or manage the use of the equipment [Regulations 8 & 9]. For example, by providing the manufacturer's instructions to operators and by sending operators on training courses if required.

Where the equipment may pose a risk to health or safety, to restrict its use and maintenance to specific persons [Regulation 7]. For example, a forklift truck operator will require a certificate of competence before he can operate the machine.

To ensure that work equipment is maintained in an efficient state, in efficient working order and in good repair (service and maintenance records kept as evidence) [Regulation 6]. For example, by having a maintenance contract for some items of equipment or by regular scheduled servicing.

To provide adequate protection from dangerous parts of machinery and rotating stock bars [Regulation 11]. For example, by the fitting of machine guards.

To take measures to prevent or control hazards created by articles or substances falling or being ejected from work equipment, or by discharge from the work equipment, fire or explosion [Regulation 12].

To provide protection to prevent injury from burns, scalds, and searing from high or low temperature work equipment, parts of work equipment, and articles and substances produced, used or stored [Regulation 13].

To provide controls and control systems for work equipment, for example, stop controls and the positioning of these controls [Regulations 14 to 18].

To ensure isolation from sources of energy [Regulation 19].

To take measures to stabilise work equipment [Regulation 20].

To provide suitable and sufficient lighting [Regulation 21].

To take suitable measures to ensure that work equipment can be maintained whilst shut down [Regulation 22].

To clearly mark work equipment with Health & Safety warnings [Regulation 23].

A copy of any manufacturer's instructions should be kept available for the users of any work equipment and an inventory of all work equipment should also be kept. The latter may also be a requirement of your insurance policy as well as being required for your financial auditing and records.

Regulation 26 requires any work equipment that an employee rides upon (a tractor or fork-lift truck for instance) that is capable of rolling over and crushing the employee to be fitted with a device to stabilise the equipment or a structure (Roll over prevention system – ROPS) to prevent the equipment rolling over completely (only onto its side) – a structure giving sufficient clearance to anyone being carried if it overturns further than that or a device of comparable protection.

Where there is a risk of anyone being carried by mobile work equipment being crushed by its rolling over, the employer shall ensure that it has a suitable restraining system for him.

To sum up, there should be no risk to health and safety from the use of any work equipment regardless of age, origin or its condition.

All work equipment must be tested or inspected at the required intervals to ensure it is safe to use. All workers must be adequately trained and 'competent'. For example, no one should ride on mobile work equipment (including any vehicles) unless the equipment is suitable (designed, equipped and capable of carrying passengers) and risks are minimised.

Lift Trucks

All mobile plant equipment such as forklift trucks, dumpers and tractors which have been provided for use before 5th December 1998 should have been brought into compliance with Part 3 of the Provision and Use of Work Equipment Regulations 1998 (PUWER 98) by 5th December 2002.

In real terms, this has a major impact on employers, the self-employed or anyone in control 'to any extent' of mobile work equipment. It means that all relevant parties must ensure that their equipment is safe to use by preventing or controlling risks. This is not a completely new concept as the requirement of the regulations has been in effect for new equipment since the 5th December 1998.

There are many risks involved in the use of mobile work equipment. Some of the most common risks are: being struck by the vehicle; the equipment tipping or rolling over; the operator of the equipment falling from the vehicle; the operator being struck by falling objects; unauthorised operation and use of the vehicle by untrained persons; braking systems not being maintained to correct and safe standards; and bad or restricted driver visibility.

One simple means of controlling the risk is by restricting the use of the equipment to only authorised users; this is often easily achieved by only allowing the access of the 'key' to the relevant authorised persons. It is suggested that a monitoring and control system be established such as the signing out and in of keys.

PUWER 98 addresses many of the risks in Part 3 by insisting that employers now take adequate measures to ensure that all work equipment, in particular plant equipment, is safe to use.

There are many ways in which this can be achieved but one of the main areas to be addressed is the fitting of restraining systems (seatbelts, lap-belts, etc) – PUWER 98 states under Regulation 27 the requirement for the provision of restraining systems to prevent crushing of the operator between the truck and the ground where there is a foreseeable risk of overturning. Due to the nature of a lift truck, all situations during use could constitute a risk of overturning, therefore all forklift trucks must be fitted with restraining devices of some kind by 5th December 2002. Further information on 'Fitting and use of restraining systems on lift trucks' can be obtained in the HSE information sheet MISC 241). It is a fact that unsafe work equipment is a major factor in a very large percentage of deaths and injuries every year. The statistics for 2001 alone show that three people were killed and 67 were seriously injured when they were struck by lift trucks.

10 THE SUPPLY OF MACHINERY (SAFETY) REGULATIONS 1992

These regulations apply to those who supply or manufacture machinery. The regulations and schedules are extensive so I will not go into detail here and will instead only refer the interested reader to the Regulations for full information. From 1st January 1993, these regulations secured free access for machinery manufactured or imported to the United Kingdom and throughout the European Community (EC). To comply, the machinery shall comply with safety requirements illustrated in Schedule 4 of the Regulations and carry the European Community CE mark of approval.

Refusal of access of CE approved machinery to other member states is *not permitted.* These regulations are applicable to all forms of machinery, other than those specifically excluded.

Suppliers shall, through certification, ensure that a machine satisfies essential Health & Safety requirements. The supplier is required to retain a file and provide information upon request and this must include sufficient technical information clarifying how the machine satisfies requirements.

Where machinery is deemed specifically hazardous, a technical information file shall be made available by a *test house*, demonstrating the machine's compliance with EC safety standards. Where no standard relevant to the machine exists, or the machine fails to conform to a known standard, the machine shall be tested, thereby ensuring compliance with safety standard requirements.

The definition of the word 'safe' within the regulations means the machine, when installed, maintained and used in accordance with the original design intent, is 'Fit For Purpose' (FFP) and there being, so far as is practical, no risk to persons, property or domestic animals.

11 SELECTION AND MANAGEMENT OF STAFF AND CONTRACTORS

It is in this area that quality, competence and Health and Safety are linked most closely. The law requires you to use investigative means to ensure the competence of the staff and contractors you intend to appoint. The advice is to have a written set of procedures for selecting staff, contractors and suppliers.

For selecting staff, this may include one or more of the following techniques:

- Formal and informal interview
- Checks on records of training, experience and qualifications, including Safety Passport schemes
- CVs and references
- Recommendation and reputation
- Observation and previous experience of working with that person
- Health and fitness checks if these are felt necessary, or the nature of the work dictates.

Advice on fitness and health checks can be obtained from the HSE Employment Medical Advisory Service (EMAS).

It is an employer's responsibility to ensure that information and records for individuals are kept in confidential personnel files, remembering also to comply with the Data Protection Act 1998. As a minimum, the information required for each individual should includc:

Name
Address
Date of Birth
National Insurance
Nationality

A Passport or Driving Licence should also be used to confirm an employee's identity. Don't be surprised if one day the police ask for these details about your employees; anti-terrorist squad officers have warned that foreign nationals involved with terrorist or protest activities are attempting to access high-profile events by posing as workers. This is a problem that may affect work agencies or those organisations with high numbers of temporary staff such as stewarding, cleaning and crewing companies. Anti-terrorist squad officers are using the Asylum and Immigration Act 1996 as a means of disrupting and discouraging terrorists, political activists and protesters from gaining access by means of employment to high-profile events.

When selecting contractors for their lists of Approved Contractors, many Local Authorities use a system or series of checks that were originally designed to cover the CDM Regulations.

To even get a chance to supply a quote to many Local Authorities or large organisations, a company may need to be on this list of Approved Contractors. To get listed you will be required to supply a great deal of information about your company's H&S management systems (that will then be thoroughly scrutinised and checked, often against a 'points' scoring system) before you finally get on one of these lists.

One reason why Safety Passport schemes are being introduced in many industries is to assist contractors in getting onto Approved Contractor lists.

The kind of information normally required when selecting contractors to ensure they are 'competent' to fulfil their Health & Safety responsibilities is as follows:

- Accident Records (Including RIDDOR) and Reporting System
- Risk Assessments
- Health & Safety Policy Statements *(In addition to the basic Policy Statement, this document must include information on the management structure and organisation in respect of health and safety and the Health & Safety arrangements the company/business has in place.)*
- Method Statements and Safe Systems of work
- Details and Records of Staff Training and Training Schemes operated
- Qualifications
- Test/Inspection/Examination Certificates or records for equipment including fire, work at height and LOLER.

All this seems a lot to ask for, but it is now quite normal within most industries to ask for this kind of evidence, we are just going to have to get to grips with it.

Should a supplier/contractor have a serious accident at the site/venue that a production manager/employer is in charge of, the production manager/employer could be in serious trouble if he/she has not obtained copies of, checked and recorded, this information from the contractor in question.

All these checks are part of your 'quality' management system: *you are responsible for the contractors you appoint* and unless you have checked their competence then there is no way you can call your business a 'quality' business.

You can't just appoint contractors and 'let them get on with it'. If your contractors fall foul of Health & Safety laws while working on your show or event then you may well be held responsible and can be prosecuted.

A tip for those appointing contractors: always ask a contractor to provide a Health & Safety Policy, Risk Assessments, Insurance information etc. with the quote. Select contractors not just by their quote, but by a combination of quote price and suitable and sufficient Health & Safety information. It's no good asking for the information after you have confirmed the contract, you'll be lucky to get the information and you won't know if you're appointing competent contractors if you haven't seen their Health & Safety documentation before you appoint them.

You need this information *before* you appoint them because you must ensure the people you intend to appoint (not the people you have already appointed) are 'competent'. If you appoint them before checking they are competent you've obviously got it wrong and may be breaking the law.

While many companies and employers are obtaining all the required Health & Safety information from the contractors they are about to appoint, they seem to forget that the reason they have obtained this information is so that they can check to see if the contractor is 'competent'. In other words, he has suitable and sufficient safety control systems in place, and of course this can't be done if the employer or person engaging the services of a contractor does not have sufficient knowledge of the job/system or process and sufficient Health & Safety knowledge to be able to make this assessment.

All too often I have seen promoters and production managers with many safety policies and risk assessments from contractors that have never been read and even if they did read them they would probably not understand them; this is where the use of competent persons who have the required experience and knowledge of the job, work system or process as well as Health & Safety experience, is again required to advise on such Health & Safety matters. If, after checking to see if the proposed control systems are adequate, anything is found lacking, the contractor should be informed and asked what he proposes to do to control the risk and for a new risk assessment to be provided; any contractors who are unable or unwilling to do this must be rejected.

To sum up, it's a total waste of everyone's time to ask for suppliers' Health & Safety policies and risk assessments and just file them away, you need to be 'competent' or have access to 'competent' persons that can advise you so that you are able to act upon the information you have gathered to ensure your contractors are correctly vetted.

While not a legal requirement, it is considered 'good practice' to have a Health & Safety Policy for contractors that clearly sets out what is required of contractors in terms of Health & Safety management; this may be part of your main company's Health & Safety policy or a separate document, but it obviously needs to be issued to and brought to the attention all contractors and suppliers to be effective.

Having appointed 'competent' contractors and/or staff it is still your responsibility to monitor their activities on site to ensure compliance and to take action if the contractors or staff are not complying. It is increasingly common on festival sites and in high profile venues to have a safety adviser or consultant on hand to carry out continuous monitoring of contractors and who may be responsible for ensuring the required information is supplied prior to appointment. Unfortunately, some of these consultants/advisers do not act on the information provided by suppliers or contractors and simply file it away, therefore they do not do a sufficient job in trying to legally protect their clients or establish safe working practices and/or environments. In some cases this may be even worse because the consultant/adviser that has been appointed is not conversant with current industry best practice, techniques or regulations. The term 'chocolate tea pot' springs to mind in this situation!

All contractors must be informed that their Health & Safety control, management systems and performance will be monitored and measured, and that financial penalties and/or disciplinary action will be introduced or taken if the minimum statutory requirements are not met by contractors.

Employees must also be monitored and additional training, supervision or information be provided if necessary.

The monitoring of contractors' level of Health & Safety compliance is a vital function of Health & Safety advisers and consultants. A Health & Safety consultant must have access to contractor's assessments and policies etc. Contractors must be informed that these monitoring procedures will take place and of the systems in place to deal with those whose Health & Safety systems fail to meet the minimum statutory requirements.

12 TRAINING

The Heath and Safety at Work Act requires employers to ensure adequate training is given to employees. The Management of Health and Safety at Work Regulations require employers to train their employees adequately at the recruitment/induction stage, before exposure to any new or increased risk, or when transferring them to work using different equipment or processes. As well as the Health and Safety at Work Act, many of the specific regulations that fall under the Act require training to be provided as a means of controlling safety. They include:

The Manual Handling Regulations
The Noise at Work Regulations
The Display Screen Equipment Regulations
The Provision and Use of Work Equipment Regulations
The Health and Safety (First Aid) Regulations
The Personal Protective Equipment Regulations
The Lifting Equipment and Lifting Operations Regulations
The Management of Health and Safety at Work Regulations
The Work at Height Regulations

The list is not exhaustive and the training should include:

Fire Safety and emergency procedures
Fire Aid provision
Accident reporting
All aspects of Health and Safety involved with new work equipment or processes
The limitations of use of work equipment
Inspection, testing of work equipment
The use of work equipment
The problems that could occur in using the work equipment.

How to deal with these difficulties:

Making use of any experience in using the equipment to predict and control risk.

Making everyone aware of the risks and control measures in place

The training should take account of the individual capabilities of the people involved and be repeated enough to remain effective.

The training must be carried out during normal working time and trainees must be paid their normal daily rate of pay during training and, where necessary, account must be taken of specific regulations or requirements of certain industries. The self-employed have to arrange for the provision of and pay for their own training.

Records of training should be kept by the employer and show the names of those trained, who trained them, what was included in the training and the times and dates of training.

The objectives, content and results of training as well as source, qualifications and validity of the course(s) should be recorded and a review date set to allow the standard and benefits of the training to be maintained.

The Benefits of Training

- Training is a requirement of most safety legislation, HASAWA, MHASAW, PUWER, LOLER, PPE and Manual Handling Regulations etc.
- Training is a means of achieving competence.
- Converting information into safe working practices contributing to a safety culture in the workplace.
- A means of increasing safety and productivity
- Training Needs

  Can be determined by risk assessment, e.g.:

    Basic Skilling

    Specific on the job training

    Health & Safety training

    Emergency procedures training
- Training is necessary when:
- New staff are recruited – induction training
- Changes occur in the work activity
- Changes occur in the work environment
- Changes occur in the work equipment or working methods
- Refresher training is required by all

  - Skills decline with time and need practice

  - Personnel who dispute with other employees

- Young people need special training
- They lack 'life' skills and experience
- They require greater supervision

Before safety-training programmes can be constructed, an assessment of the need for safety training should be conducted to ensure that the precise need, target group and content are properly identified, in other words, a *training needs analysis* should be undertaken. Indicators of a need for training include:

- A high or increasing accident rate by comparison with similar companies
- Excessively high employees' liability insurance premiums
- Excessive waste or scrap from machines or processes
- Undue attention being paid to the company by enforcement officers
- A high rate of property damage
- Poor quality work

Basic or induction training may be carried out 'in house' by a 'competent person' or a specialist consultant or training provider may be engaged to provide the training.

Many companies with a high turnover of casual or part-time staff, such as crewing or stewarding companies, don't see the value of training because they feel that staff are not with them very long and the costs would be prohibitive. These companies should consider offering training that is of real value, as it is very likely that there would be a better retention of staff and that staff turnover would be reduced, the staff remaining would also be safer and of a better quality.

Safety Passports

Many industries, large companies or groups of companies have now adopted Safety Training Passport schemes, the principal of such schemes is fully supported by the HSC, the Environment Agency and the HSE who have issued guidance on there content and implementation.

A Health & Safety training passport is similar to any other passport; it allows the holder access to passport-controlled work environments. They are designed to be used by clients seeking assurance on levels of contractors' Health & Safety awareness. Like induction training, safety-training passports offer basic, practical Health & Safety training tailored to suit the industry. Successful trainees are usually issued with a credit card sized I.D. card that is secure and easily verified from a central database.

Safety Passport Benefits

- Safety Passports are a very simple way for workers (including the self-employed) who move from one contract or company to another to show employers they have received basic training.
- Safety Passports save time and money because workers need less induction training.
- They reduce accidents and ill health at work.
- They can have significant impact on reducing pollution incidents, minimising waste and contribution to a cleaner environment for everyone.
- Companies know that workers have been trained to a common, recognised and validated standard.
- They show a companies commitment to having safe and healthy workers.
- They help promote good practice in the supply chain between contractors and companies.
- Insurance and liability premiums may be reduced if a company can show that all workers have basic Health & Safety training.

Safety Passports are not:

- A way of knowing a worker is competent
- A substitute for Risk Assessment
- A way of showing 'approval' of a contractor
- Required or regulated by law
- A reason to ignore giving specific information or
- A substitute to effective on-site management

Syllabus for Safety Passport or Induction Training

The basic syllabus for Safety Passport or Induction Training should include:

- An introduction to Health & Safety Law
- Workplace Safety
- Fire Precautions and Procedures
- First Aid and Accident Reporting
- Manual Handling
- Personal Protective Equipment
- Electrical Hazards
- Safety Signs

- Plant and Machinery including Workplace Transport
- Safe Systems of Work

Additional training may be required for those working at height, users of hazardous substances or Display Screen Equipment and those exposed to high sound pressure levels.

More skill specific training (if not available in house) may be available from some equipment manufactures or suppliers but probably the best skill specific training will be the in the form of nationally recognised qualifications such as BTEC, City and Guild, NCFE or NVQ qualifications.

Several attempts have been made to introduce formal training and qualifications of the types mentioned above into our industry with varying degrees of success, including a major scheme by the Production Services Association to develop and provide BTEC qualifications.

A great deal of training and qualifications already exists, and drivers, electricians, first aiders, plant operators and caterers are well covered. Recently, a number of companies that provide rigging courses on a commercial basis have got together to produce a National Rigging Certification scheme that is getting major support from the industry and the stewarding sector is now well on course with its training programme under the wing of the UK Crowd Management Association.

Several important areas sadly neglected in training programmes include the local crew, stagehands and those entering the industry. A Safety Passport scheme would be ideal not only for them, but all who work in our line of work.

It's not only the lower ranks that are neglected, but also the likes of Production Managers and Tour Managers, who need not only high levels of experience but also some kind of bespoke formal qualification with an emphasis on Health & Safety.

While on this subject, I must point out the difference between training, instruction and information. Information is what it says, and having been given information one would hope that the person receiving the information has the ability to act accordingly.

If you were to tell someone: "Stay calm if a fire breaks out", that would be giving that person an instruction. But if you had explained and perhaps demonstrated what to do and 'how' to stay calm if a fire broke out, that would be training.

One quite obvious thing about training is that there must be a standard or benchmark to train to.

In this respect, it's a pity that the *Event Safety Guide* is only guidance and not an Approved Code of Practice. If it were, then it could be the minimum 'benchmark' standard to use when we are developing our training and qualifications.

It has long been agreed that training holds many benefits, but unfortunately in our industry not many people bother to undergo any kind of formal training and therefore do not reap these benefits. No matter how well trained individual staff may be, they cannot operate safely unless the whole working environment has a suitable safety management system in place. For events, the responsibility of putting a safety management system in place will normally fall to the Production Manager. The Production Manager must work as a team with the Event Safety Officer/Advisor and the Crowd Safety Manager, who must not only have knowledge of Health & Safety/crowd management, but also a comprehensive underpinning knowledge of the production equipment, systems and techniques used in each area.

13 HEALTH & SAFETY POLICIES

Any company or business that employs five or more people is required under the HASAWA to have a written policy on Health & Safety; it is also considered good practice for a business with fewer than five employees to have a Health & Safety Policy.

The Health & Safety policy is a real starting point, it sets out the official general policy and arrangements for the time being in force with regard to Health & Safety within that company or business for all employees, contractors and members of the public who may be affected by the actions of the company or business. The policy also sets out the management organisation in respect of Health & Safety and the arrangements for carrying out that policy.

The policy must state with whom the overall responsibility for Health & Safety lies, as well as other responsibilities. This is normally the proprietor, managing director or CEO of the business, and he or she must also sign the policy. The regulations require all employers to bring this policy and any revisions to the notice of all of his or her employees, including part-time and casual workers so they are aware of what is expected of them.

Basic Objectives and Content of the Statement of Intent

Health & Safety policy statements should state their main objectives, e.g.:

(a) Commit to operating the business in accordance with the Health and Safety at Work Act 1974 and all applicable regulations made under the Act, 'so far as reasonably practicable'.

(b) Specify that Health & Safety are management responsibilities ranking equally with responsibilities for production, sales, costs, and similar matters.

(c) Indicate that it is the duty of management to see that everything reasonably practicable is done to prevent personal injury in the processes of production, and in the design, construction, and operation of all plant, machinery and equipment, and to maintain a safe and healthy place of work.

(d) Indicate that it is the duty of all employees to act responsibly, and to do everything they can to prevent injury to themselves and fellow workers. Although the implementation of policy is a management responsibility, it will rely heavily on the cooperation of those who actually produce the goods and do the work and that take the risks.

(e) Identify the main board director or managing board director (or directors) who have prime responsibility for Health & Safety, in order to make the commitment of the board precise, and provide points of reference for any manager who is faced with a conflict between the demands of safety and the demands of production.

(f) Be dated so as to ensure that it is periodically revised in the light of current conditions, and be signed by the chairman, managing director, chief executive, or whoever speaks for the organisation at the highest level and with the most authority on all matters of general concern.

(g) Clearly state how and by whom its operation is to be monitored.

Organisation (People and their Duties)

Suitable policies will demonstrate – both in written and diagrammatic form (where appropriate) – the following features:

(a) The unbroken and logical delegation of duties through line management and supervisors who operate where the hazards arise and the majority of the accidents occur.

(b) The identification of key personnel (by name and/or job title) who are accountable to top management for ensuring that detailed arrangements for safe working are drawn up, implemented and maintained.

(c) The definition of the roles of both line and functional management. Specific job descriptions should be formulated.

(d) The provision of adequate support for line management via relevant functional management such as safety advisers, engineers, medical advisers, designers, ergonomists, etc.

(e) The nomination of persons with the competence and authority to measure and monitor safety performance.

(f) The responsibilities of all employees.

(g) The arrangements for employee representation on Health & Safety matters (i.e. whether by trade union safety representatives, employee elected safety representatives, or by direct consultation with each employee. (See 'joint consultation', 'safety representatives' and 'safety committees').

(h) The involvement of the safety adviser and relevant line/functional management at the planning/design stage.
(i) The provision of the means to deal with failures in order to meet job requirements.
(j) The fixing of accountability for the management of Health & Safety in a similar manner to other management functions.
(k) The organisation must unambiguously indicate to the individuals exactly what they must do to fulfil their role. Thereafter a failure is a failure to manage effectively.
(l) The organisation should make it known – both in terms of time and money – what resources are available for Health & Safety. The individuals must be certain of the extent to which they are realistically supported by the policy and by the organisation needed to fulfil it.

Arrangements (Systems and Procedures)

It is vital to establish safe and healthy systems of work designed to counteract the identified risks within a business. The following aspects should be used as a guide when preparing arrangements for Health & Safety at work:

(a) The provision of Health & Safety performance criteria for articles, and product safety data for substances, prior to purchase.
(b) The provision of specific instructions for using machines, for maintaining safety systems and for the control of health hazards.
(c) The development of specific Health & Safety training for all employees.
(d) The undertaking of medical examinations and biological monitoring.
(e) The provision of suitable protective equipment/clothing.
(f) The development and utilisation of permit-to-work systems.
(g) The provision of first-aid/emergency procedures, including aspects of fire safety/prevention.
(h) The provision of written procedures in respect of contractors and visitors.
(i) The formulation of written safe systems of work for use by all levels of management and work force.

Other matters that might also be referred to include the arrangements for compliance with the Health and Safety (Display Screen Equipment) Regulations 1992, the Management of Health and Safety at Work Regulations 1992

Regulation 3 (risk assessments), and the disciplinary measures consequent upon a breach of the policy.

Appendices to Statements

There are a number of reasons for incorporating appendices to statements of Health & Safety policy (although this is not a statutory requirement). For instance, there may be a need to detail the organisation's intentions, arrangements and procedures for dealing with a hazard specific to a process, e.g. the risk of back injury associated with a particular handling operation. It may be necessary to formally declare the company's policy on asbestos in existing buildings or on the provision of prescription lens eye protection to certain groups of operators. Any company or organisation that deals with children, such as an event that admits children or a stewarding company working on an event where children are admitted, will require a Child Protection Policy. Fundamentally, an appendix qualifies in depth certain provisions outlined in the policy.

Policy monitoring

Policy monitoring highlights four areas, as follows:

(a) The accident and ill-health record.
(b) The standards of compliance with legal requirements and codes of practice.
(c) The extent to which organisations specify and achieve – within a given time scale – certain clearly defined objectives (of both short-term and long-term nature).
(d) The extent of compliance with the 'organisation' and 'arrangements' parts of the organisation's own policy (discussed earlier), including in particular the written safe systems of work that have been developed by the organisation to meet its individual needs.

Plant Equipment and substances

- Maintenance of equipment such as tools, ladders, etc.
- Are they in safe condition?
- Maintenance and proper use of safety equipment such as helmets, boots, fall arrest equipment, etc.
- Maintenance and proper use of plant, machinery and guards.
- Regular testing and maintenance of lifts, hoists, and other dangerous

machinery, emergency repair work, and safe methods of doing it.

- Maintenance of electrical installations and equipment. Safe storage, handling and, where applicable, packaging, labelling and transport of dangerous substances.
- Controls on work involving harmful substances, such as lead and asbestos.
- The introduction of new plant, equipment or substances into the workplace
  – by examination, testing and consultation with the workforce.

Employees must be aware of the policy and, in particular, must understand the arrangements which affect them and what their own responsibilities might be. They may be given their own copy (for example, within an employee handbook) or the policy might be displayed around the workplace. With regard to some arrangements detailed briefings may be necessary, for example as part of induction training. Employers must revise their policies as often 'as may be appropriate'. Larger employers are likely to need to arrange for formal review and, where necessary, for revision to take place on a regular basis (e.g. by way of an ISO 9000 procedure). Dating of the policy document is an important part of this process.

Your Policy Statement is a legal document; it may be used to show to others how you intend to manage Health & Safety, yet at the turn of a coin it may also be used against you in a Court of Law if things go 'apex up' and it can be demonstrated to the Court that you are not following the Policy and safety procedures that you have produced. When you write your Safety Policy you may be making a rod for your own back.

Promoters should prepare a Health & Safety Policy, even when they employ fewer then five persons. This should state their policy in respect of the overall safety of the event and cover such topics as that of public safety that are within the organisers' control. The policy should also set down standards of safety expected from contractors, caterers and others.

Whilst not a legal requirement, it is considered 'good practice' for all businesses to have a Health & Safety policy for contracts that clearly sets out what is required of contractors in terms of Health & Safety management. This may be part of your main company or a separate document that should be issued to all contractors.

A Health & Safety policy and most Health & Safety management work is all about **PETER, P**ersons, **E**quipment, **T**asks, **E**nvironment and **R**eporting (or

Recording) and must be reviewed when any of these changes.

Of course, not all the dangers and hazards may be created by your operations; often they are created by others working alongside.

Employers and those in charge have a duty to inform and warn other workers of these dangers and the safety systems that should be in place to control these dangers (Regulation 12, Management of Health and Safety at Work Regulations 1999).

For example, a rigger may drop a lamp from out of the truss onto the head of a backline technician, or the PA may get turned full on when you are standing two feet in front of the bass bins and your hearing is damaged. You should have been warned of these dangers and records kept to prove that you had received the warning.

Like it or not, this is where some kind of training backed up by certificates of competence are so important and is also a very good reason why it is worth developing standards and qualifications that are specific to our industry. They show you have received training and information and that you are aware of your legal responsibilities as well as showing competence in the tasks you've been asked to undertake. Training is one of the main systems for controlling safety and establishing 'safe systems of work'.

14 RISK ASSESSMENT

Risk Assessments now form much of the basis of modern Health & Safety thinking and approach and they are used to help create safe systems of work. There has been a legal requirement to carry out risk assessment and act on the conclusions for over 20 years. The Management of Health and Safety at Work Regulations 1999 details requirements for risk assessments.

- Regulation 3 requires the company to complete assessments on all its activities and record the significant findings.
- Regulation 8 requires that employees be provided with relevant and suitable information on any risks to themselves.

At its most basic, a Risk Assessment is an analysis of how dangerous a certain situation or activity is, what the likelihood is of an accident occurring and what the implications of such an occurrence are. Old-style Health & Safety legislation set out a list of 'dos' and 'don'ts' for set situations and hazards. This was a prescriptive and very inflexible approach and did not allow for many of the work situations, environments and operations carried out these days or the number of people who may be affected. The new style of assessment gives management the responsibility to assess not just hazards and risks and then taking what ever action is called for by the assessment, but also the responsibility to apply the same process to assess other needs such as what First Aid or Fire Safety facilities may be required at a certain work place.

Risk assessment identifies hazards (something with the potential to cause harm) and identifies the level of risk (the likelihood of harm and seriousness of its consequences).

The process continues with the implementation of precautions to eliminate or reduce the risk to an acceptable level. Monitoring operational risks and introducing further controls if necessary are the final steps in risk assessment.

Remember - assessment is not an end in itself; it is merely a structured way of analysing risk and pointing the way to practical solutions.

We live in an age when parents can sue a theatre for scaring their child with a production of *Peter Pan* – what are the legal implications for an employer whose workers may be injured by an accident about which they had no prior warning, or which they were led to believe was perfectly safe?

It's a good job that child had not been to an Ozzy Osborne or Alice Cooper gig! The courts have had to decide on civil claims arising from injuries allegedly sustained through use of a *typewriter*, just think of the possibilities in our industry! Compensation awards are retrospective: the courts have awarded a man working as a 'digger driver' between 1987 and 1990 £187,000 against his old employer for Repetitive Strain Injury (R.S.I.) gained from driving the digger.

At this point I should also like to quote a story told to me by crowd safety supremo Mick Upton that concerned an insurance claim against Rod Stewart.

It transpired that Mr Stewart liked to kick footballs from the stage into the audience at the start of his concerts and at one concert a gentleman raised his hand into the air above his head in order to catch one of these footballs. Now, this gentleman was obviously not a goalkeeper as the ball missed the gentleman's hand but did manage to strike the top of his right index finger bending it backwards and breaking his finger.

The gentleman in question successfully sued Mr. Rod Stewart for breaking his finger and ruining his sex life...

I laughed as well when I heard that story, but honestly it is true and just shows how important it is to risk assess the artist's performance. Kicking footballs into the audience or throwing objects into the audience that will cause a crowd serge, jumping off stage into the audience, inciting the crowd to behave in a dangerous or anti-social manner – these possibilities must all be considered and assessed.

Quite surprisingly, the purpose of risk assessment is not directly to help prevent accidents and reduce risks, rather it does this by default. The Approved Code of Practice to the Management of Health and Safety at Work Regulations states "*the purpose of Risk Assessment is to help the employer or self-employed person to determine what measures should be taken to comply with the employer's or self-employed person's duties under the 'relevant statutory provisions'*." By complying with the relevant statutory provisions, risk will have been reduced.

The most common mistake that is made with risk assessment is failing to check if the systems and methods introduced to control risk meet with the legal requirements and standards as shown in HSE Guidance and Approved Codes of Practice. It is vital that Guidance and ACoPs are referred to by the person making the assessment and are then listed as references in the assessment.

As mentioned, risk assessment is now a significant part of a risk management system.

There is a rumour that in the near future it is likely that all businesses (regardless of the number of employees) will have to produce written risk assessments and a written Health & Safety Policy. The reason for this is that the HSE is very aware of the number of small businesses which are having accidents during dangerous work operations yet do little if anything in terms of creating safe systems of work, hiding behind the fact that they are not required to produce written assessments and policies. It's only a rumour at present but I can see the logic behind it all.

The Risk Assessment Process

All of us carry out risk assessments every day without even thinking about it, such as when we cross a busy road or drive a vehicle, but for our purposes of establishing a 'safety culture' we must adopt a more formal approach and record the assessment in writing.

It is the assessment process that most people get wrong. For example, I have seen assessments for Manual Handling that state: *All crew are experienced and therefore our assessment is complete and we need do no more.*

Wrong! This is an insufficient and unsuitable assessment and will not be acceptable to the HSE or get you out of trouble when things hit the fan. There is a lot more to be done as the Management of the Health and Safety at Work Regulations require an assessment to be suitable and sufficient to reduce the risk to the lowest possible level and the example we have seen here clearly does not achieve this.

In fact it may even be better to have no risk assessment, then an assessment like the above example as the law will probably come down on you harder for only playing lip service to your Health & Safety legal responsibilities.

No mention has been made in the assessment of the regulations or about other methods of controlling the risk by either removing the hazard or, if this is not possible, other means such as training and as a last resort, protective clothing such as safety boots, and while the crew may be experienced they still may not be doing it safely. Obviously, the person making the assessment had not done any research and found out what the regulations required (see Manual Handling Regulations for further details). *Remember, the assessment has to fit the task: don't try to fit the task to the assessment!*

The person carrying out a risk assessment must be competent for the task. In this context competency means:

- understanding current legislation and best practice

- knowing one's own limitations of experience and knowledge
- willingness and ability to increase experience and knowledge

The hazards we are assessing need not be a direct result of your operation, they may be due to other contractors or inherent environmental risks – either way the employer/contractor has a duty to identify any *significant* hazards, to carry out an assessment, to record the risk assessment in writing and inform his/her staff. This is a requirement under Regulation 3 of the Management of Health and Safety at Work Regulations 1999.

There are many different types of forms available for recording risk assessments including forms for specific types of assessment such as DSE or Fire Safety, or you may wish to use your own system. In an emergency, even making a quick assessment record on the 'back of a cigarette packet' is quite acceptable; it at least shows you were trying and adapting to changes in circumstances site and conditions as they arose. Whatever method is chosen, the following points marked *** must be included:

Date Of Assessment

Name The Person Making The Assessment

Name the job or operation to be assessed (if applicable, if not, give a brief job description)

Identify the Hazards ***

These hazards may include:

- Fire
- Chemicals
- Vehicles or plant
- Moving machinery parts
- Falls (falls from heights, falls at the same level and falls into holes)
- Noise
- Poor light
- Slips and trips
- Weather
- Manual handling
- Electricity
- Work Hours
- Stress

- Environmental hazards (Cliffs, rivers, streams, ponds, the sea, ditches, hedges, fences, walls and street furniture)
- Biological Hazards (Tetanus, Lime Disease, E.Coli and Leptospirosis as well as the hazards caused by plants, insects and vermin)
- Litter and Waste products
- Aggressive or violent people
- Crowds

You will almost certainly find there is more than one hazard associated with each work activity. Prioritise and list them in order of highest risk, then it will be possible to carry out a separate assessment for each hazard you have found starting with the most obvious or serious risk. It is a common mistake to try and assess all the hazards in one assessment; it's risk assessments not risk assessment!

List those at risk and approximate numbers *** (List these as groups not by individual name)

These may include:

- Your own staff
- Other contractors and performers
- Stewards
- Members of the public

What is the likelihood of an accident?

Impossible, remote, possible, probable or likely?

(Consider what the likelihood will be *without any* controls in place whatsoever, don't make the mistake of assuming that some controls are already in place).

What is the worst possible outcome of an accident?

(Consider what the outcome would be *without any* controls in place whatsoever, and don't make the mistake of assuming that some controls are already in place).

It may be:

- Equipment Damage - No Injury
- Trivial Injury
- Minor Injury
- Major Injury
- Fatal Injury

Risk Class

The above information can now help us to class the risk as High, Moderate, Minor or Acceptable.

- Risk = Hazard Severity x Likelihood of Occurrence (x Number of People Exposed)

A Hazard Index Table can be produced:

	Fatal 5	Major 4	Minor 3	Trivial 2	Equipment Damage 1	
Likely 5	[illegible]	20	15	10	5	1 in 10
Probable 4	[illegible]	[illegible]	12	8	4	1 in 100
Possible 3	15	12	9	6	3	1 in 1000
Remote 2	10	8	6	4	2	1 in 10, 000
Improbable 1	5	4	3	2	1	1 in 100, 000

Having classed the risk we can now look at ways of controlling the risk.

Information

List here the current information available on the hazards, you will have to do some research to obtain the information required that will help identify the appropriate methods of controlling the risks. Remember, *your controls must meet the standards set by legal requirement* so you will have to research documents and publications including:

- Approved Codes of Practice
- HSE Guidance
- Safety Data Sheets
- Statutory Regulations
- Manufacturer's Instructions
- Company Safety Rules or Policy
- British/GEN Standards

- Codes of Practice
- Industry Best Practice
- Personal Experience

This may seem an incredible amount of work but it's not really that bad, most of the documents and publications you need to research are listed at the back of this book or referenced in the various chapters and many of the basic documents are summarised within this book.

Current Controls ***

What are your present systems for controlling the risk? Controls will fit into one of the following groups:

- Architectural
- Managerial or
- Physical

An example of an architectural control is the safety refuge in a loading bay or constructing a building from stone, concrete and metal (instead of wood) to prevent fire, an example of a managerial control is the employer or production manager giving an instruction or warnings or it may be a safety sign and a physical control maybe something like a portable loading ramp.

Do your current controls meet the standards set by legal requirement? What is in place right now? *Do not list what you intend to do in the future, only what is in place now.*

The next common mistake is to assume that some things are already in place to control the risk; you probably have little on nothing in place right now.

These may include any of the following controls, remember, the first step in controlling a risk is to try to remove the hazard.

The Hierarchy of Controls Systems:

- Remove the hazard (the best option if it's possible to do so)
- Access controls to the work area
- Training, Qualifications, Supervision and Instruction
- Written safe systems of work such as Method Statements and Permits to Work
- Provision of suitable work equipment
- Testing, inspecting and certification of plant and equipment
- Structural calculations
- Warning Signs

- Personal protective equipment or clothing (very much a last resort)
- Examples of how you can remove a hazard are as follows:
  - by replacing a hazardous chemical with a safer one,
  - by engineering controls such as putting a generator in a sound proofed container to remove a noise hazard
  - by replacing a dangerous operation with a safer option
  - by mechanising a manual handling operation
  - by routing leads and cables to prevent a trip hazard
  - separation of vehicles, plant and machines from people
  - by filling in or covering a hole in the ground
  - by turning off the power before working on electrical equipment.

Are The Current Controls Adequate?

Yes or No? If the answer is Yes, move on to the section on *Monitoring*

If the answer is No:

What Further Action Is Required to Reduce The Risk To An Acceptable Level? ***

List what you need to do to reduce the risk to an acceptable level. The law says you should first of all try to remove the risk but if this is not possible then the next best option should be used. If the Risk is classed as 'High' you ***can't*** take cost into consideration when considering what control systems may be suitable.

Always remember: if at all possible our first option as a 'control system' is to always try and remove the hazard or replace it with a safer alternative and the last option we should consider is the use of Personal Protective Equipment (and clothing).

Give yourself a target date to put any new controls into practice and name the person responsible for doing this. Record all this information on the assessment sheet.

Remember, *Risk Assessment is about what is happening* in the workplace *not what you think is happening.*

This now poses a major dilemma for all involved in work at height. Anyone climbing is 'High Risk' so cost can't come into consideration when looking at control systems. We can't always remove the need to rig points or carry out other work at height.

Therefore the next best option is to use a cherry picker, lift platform, access tower, ladder, or some other 'safer' form of access system. We can only consider 'climbing' as a last resort.

But who is going to pay for the extra cost involved not to mention the extra time? More to the point, how do we get them to pay? Answers on a postcard please.

In practice, climbing will take place. On the Risk Assessment forms under 'Information' or 'Current Controls' one of the following terms is often used:

- Best Industry Practice
- Custom and Practice
- Best Practicable Means

These terms are fine on a risk assessment form but a Court may not consider them to be 'suitable and adequate' control systems for the risk involved.

Monitoring ***

This is one of the most important parts of the process. You must monitor your control systems to see that they are working adequately to control the risks. If they are not working you should review your Assessment.

Name the person responsible for monitoring the controls and reviewing the Assessment. Note how frequently you are going to monitor, list the dates or even the times.

Some operations will need to be monitored more frequently than others but assessments should always be reviewed annually or when work, materials, practices or circumstances otherwise change.

The whole point about these assessments is that they should not just be a 'paper exercise'.

To be effective all staff must be involved and the controls continually monitored to see that they are working; this is part of our 'Health & Safety Culture' in action.

When starting a Risk Assessment get the people who will actually be doing the job involved with doing the Assessment; that way they will understand the need and reason. They may be able to offer ideas and information you may have missed. Give a copy of the finished assessment to these people so that they are aware of the controls in place and can play an active role in the monitoring and controlling.

In the unfortunate event of a serious accident, the first item an HSE Officer, Insurance Investigator or Court of Law will probably want to see are copies of the risk assessment for the operation. They will then investigate to see which of the controls failed and decide who may be responsible or to blame.

As mentioned, a lot of people go wrong with risk assessments by trying to cover an operation such as unloading and rigging a PA system in one risk assessment. It may need several assessments to cover one work activity, unloading the truck, manual handling, electrical safety, work at height, rigging, trip hazards from cables and leads – that's six separate risk assessments for a start!

Employers only need to produce *written* risk assessments if they employ five or more people. However, if a significant risk could be created to a large number of people, proof of a safe system of work and training would be required by smaller employers or the self employed under The Health and Safety at Work Act 1974. Self-employed riggers may wish to make a special note of this requirement!

This does not mean you do not have to carry out risk assessments if you employ fewer than five people; you still have to carry them out but you are not required to write them down. The catch then is if you don't write them down, how are you going to prove you carried out the assessment?

Once the assessment is completed everyone affected by the work (including contractors and the self-employed) must have information about the significant results of risk assessments, including any protective or control measures introduced.

Best Practice

From time to time you may come across hazards that are not covered by specific legislation and regulation. On these occasions there are no standards to meet that are set by legal requirement, instead the controls introduced by the assessment must be generally accepted industrial standards that represent good practice and reduce risk 'so far as is reasonably practicable'. This is known as working to 'best practice' as opposed to legislation.

Generic Risk Assessments

Certain tasks that are regularly repeated may be covered by a 'generic' risk assessment. Generic assessments are perfectly acceptable but research by the HSE has found that one of thebiggest pitfalls with the risk assessment process comes about when generic assessments are used when a 'site specific' assessment is required. Users of generic assessments must have the ability to revise them to suit an individual site, task and working conditions. This means staff must know how to monitor, reassess the situation, alter the paperwork and inform all those who may be effected by the changes.

Always check that your assessment meets the minimum legal requirements and standards - this involves you finding the standards and requirements contained in HSE guidance documents and ACoPs.

If a hazard is not covered by specific regulations, any controls that are introduced must meet generally accepted industrial standards, good working practice and reduce risk 'so far as is reasonably practicable'.

15 SAFE SYSTEMS OF WORK

Having done our training, planning, and produced a risk assessment for the operation we should now have a safe system of work. Or do we?

Can we now erect a few warning signs, put on our protective clothing/ equipment (if required) and start work?

Most safe systems of work will involve keeping anyone not directly involved in hazardous operations out of the work area, and for us the work area on site is usually a stage and backstage area (indoor or outdoor) and in the back of trucks.

We need to establish traffic routes to keep pedestrians and vehicles apart, and we can keep unauthorised persons out with the use of fences, tapes, warning signs, marking off, pass systems and by the use of stewards. At the same time 'Hard Hat Areas' should be established if required.

So it's 'goodbye' to agents, managers, press, pluggers, friends, family, record company staff and liggers in general. A simple risk assessment will show that it's not safe for these people to be in our work area, because the law says we have to keep everyone safe and this is how we have to do it. It's nothing personal against these people, it's just reality! You may not be popular, but this is what must be done to control safety.

During planning and risk assessment for work on site, take a good look at the work area prior to starting work to see that there is sufficient lighting, that there are items such as hand rails where required, that it is free from all litter and spillages, and that no trip hazards exist. In other words, check out the venue or site (and make a record that you have done so).

When selecting staff be sure that they are fit and suited to the job, are trained (and certificated) or have been adequately instructed for the job in question. A supervisor (particularly with new or inexperienced crew) may be required or just someone in charge to be in control and give instructions if and when required.

Don't forget to check that all plant, machinery and equipment to be used is in good condition and ready for use; this may involve set inspection and recording procedures.

Finally, check to see that all crew are properly equipped with all the required

Personal Protective Equipment (PPE) or clothing and that they are using it correctly. This may include helmets, gloves, safety footwear and harnesses.

A good plan is to have a simple check list (covering all the potential hazards and all the above checks) with tick boxes and an 'action to be taken' (and by whom) section for anything not in order.

These check lists should be kept on file as evidence that you have made safety checks and taken any required action. It's quite possible that you do these checks anyway, and then all you have to do is make a simple record of them. It's just like another kind of risk assessment.

A good example of a Safe System of Work is not to have anyone working overhead when others are working on stage below.

If anyone is working overhead the area below should be designated a 'Hard Hat Area', warning signs erected, and non-essential persons excluded from the area. Those who by necessity need to remain below should wear hard hats.

A second example is to have a secure access ladder to a lighting or sound platform thus removing the need to climb any scaffolding or structure.

Method Statements

It should now be safe to proceed with the operation, following any written procedures or instructions that may exist. These are known as 'method statements', a set of 'step by step' instructions, rather like the instructions for building a model aeroplane kit, but it also lists the safety control measures that must be followed. Employers (production managers) should obtain copies of method statements from marquee/tent/staging/rigging/lighting companies and company's building structures, etc. Regular checks should then be made to see that the operation is being carried out according to these method statements.

If it is noticed that the method statement is not being followed then an explanation should be obtained for the deviation.

Having obtained a satisfactory explanation, get the supplier to put the explanation in writing. The same applies to variations on plans or technical drawings.

The method statement should be included as part of the overall safety plan, in fact a detailed production schedule that also includes safety instructions and information is a form of method statement. That way it will inform other workers of what is going on as well as informing employees of how the work should be done and the safety precautions to be taken.

Companies that supply equipment on 'dry hire' should also issue a simple method statement or set of manufacturer's instructions for the equipment they hire out. I know of one company that supplies power distribution equipment that supplies a set laminated instructions attached to all items supplied on 'dry hire' – a very good example to follow. This method statement can protect the hire company in the event of a claim after an accident; it may be possible to prove the equipment was not being used correctly, according to the instructions. Staff must always follow the method statement or the manufacturer's instructions.

Like copies of insurance, policy statements and risk assessments etc., method statements should also be kept on file. Contractors are not legally obliged to produce method statements but it is considered good practice.

A method statement should identify any hazards that may be present in the job and the safety precautions required to deal with them; it should also identify whether the work will be compatible with the activities of other contractors who may be present on site.

Permits To Work

A Permit to Work is exactly what it says it is! They are used in 'safety critical' situations and are a formal written system to ensure dangerous jobs are carried out safely.

A Permit To Work specifies what work has to be done, when and how it will be done, what precautions must be taken and the sequence that must be followed to ensure safety.

A Permit To Work is a legal document that must be signed; the signatory will hold responsibility for the permit.

As an example: where they are used by workmen digging holes, the permit says where they can and cannot dig so they don't hit electricity, gas, water or telephone lines.

They are used by people working on overhead power lines and often state the time work can proceed when the power is turned off, and by crews working on railway lines to ensure that no trains will be on the line.

Another example of when they are required is when storage tanks or pipes are being cleaned and inspected and they are often used during work operations in confined spaces. The permit states the times and dates that you may enter the tank, pipe or confined space.

The permit may take the form of a physical 'tally' or paper permit that has to pass to the operator when he starts work and back again when he finishes.

I do not know of any times when they are used in our business but they should be used when riggers are overhead and the area below needs to be clear of non-essential personnel.

The rigger is not given his Permit to Work (by the stage or production manager) until the area below is safe, and the rigger should not start to climb until he has been issued his permit. When the rigger descends and the area is safe he hands back his Permit to Work to show the area below is safe for everyone to resume as normal. If this system is used then it should be listed as a control method on the risk assessment and yet again, records kept. Signatures should be required on a Permit to Work to show when (and to whom) they were issued when the permit expired or was no longer in operation and when normal operations could resume.

Permits to Work should also be used for operations such as onsite welding, cutting or any live (hot) electrical work.

Permits to Work and other systems are not infallible, and certainly not foolproof. The reason for this is quite simple: fools are very ingenious, they will go to incredible lengths to find a way round a perfectly good system even if it takes twice as long to do the job as it would by following the system. The Piper Alpha platform disaster in the North Sea and a recent fatality following a fall in Earl's Court Arena are both attributed to a failure in the Permit to Work systems when someone tried to cut a few corners to get a job done more quickly.

Other Documentation

For events, a written procedure for dealing with potential major emergencies (Major Incident Plan) should also be prepared in conjunction with the police, fire brigade, ambulance/medical services and local authority. In some cases, I have known the local authority and emergency services to reject the idea of a Major Incident Plan and simply ask the event organiser to prepare their own Contingency Plans for any foreseeable problems that may conceivably be encountered during the event.

Contingency plans can cover every foreseeable kind of problem from adverse weather conditions, cancellation of all or part of the event, 'show stop' procedures, power failure, lost property, lost children, fire, stopping the performance, bomb threat, etc. The plans should go into detail on the procedures to be followed and name people or groups of people with specific responsibility to effect any action. Don't forget to give all your event staff and crew a copy of the contingency plans and ensure they are aware of their roles and responsibilities.

Large events may require an Event Manual. This can contain the Event Health & Safety Policy, Site Risk Assessments, Site Plans, Production Schedules, Staff and Crew Contacts' Lists (with Radio Call Signs), Contingency Plans, Major Incident Plans, Entertainment Licence conditions, and any other Health & Safety information the event staff, crew and contractors will require.

All safety information, including contractors' Health & Safety information such as policies and risk assessments, insurance documents, the Event Manual and the completion or sign-off certificates for structures and barriers, electrical installation and rigging should be held by the Event Safety Officer or Adviser.

Time to Start Work?

It should now be safe to start work, but we are forgetting one last thing: in an emergency do you know the position of the nearest telephone, first aid kit, emergency exit, fire extinguishers and first aider?

Regulations 9 and 10 of the HASAWA require employers and self-employed people working together to coordinate their activities, cooperate on Health & Safety and share information. Employers and the self-employed must provide information on risks and the control measures in place to address the risks to the employees of employers from outside undertakings (Regulation 12, Management of Health and Safety at Work Regulations 1999).

In other words, your employer or the venue you are working in should make you aware of any dangers that may exist (such as that hidden mineshaft on the outdoor festival site) as well as the position of emergency and first aid equipment, details of emergency procedures and evacuation, etc. At the same time you should be advised of the position of toilets, washing and other welfare facilities that may exist.

All this sounds as if it will take forever and the job will not get done but in practice it only takes a couple of minutes and is nothing to get wound up about, indeed much of it can and should be sorted out well in advance.

We Don't Need Helmets And Boots… Or Do We?

I can already hear the moans and groans about hard hats and protective footwear. Nowhere in the HASAWA does it say, "Thou shalt wear a hard hat and steel toe-capped boots" (apart from in the Construction (Head Protection) Regulations 1989).

But what it does say is that everything that is 'reasonable and practicable' must be done to make the job safer and to protect yourself and others.

Using such equipment *is* practical and not unreasonable, so there is no excuse for not using it. It is a legal requirement NOW; it is not a new regulation that may come into force in the distant future.

If you are an employer, your staff will need to be supplied with hard hats and safety footwear if risk assessment shows it to be necessary.

I have seen risk assessments that state helmets are dangerous for certain operations and therefore do not need to be worn – wrong!

If that's your view then you are obviously using a helmet that is inappropriate for the work intended, so find a different type of helmet. The law requires PPE (personal protection equipment) to be appropriate for the work.

The use of PPE is, of course, a last resort and we should always try to remove or reduce the risk so that it is not necessary to fall back on the use of PPE as a safety control method.

The self-employed must supply their own PPE. Employers should check that PPE is being used, which again will often be down to the production manager.

If it is not used, it could be very painful and costly, you may end up in hospital for a long period and not working, then being prosecuted as well as possibly giving evidence in a Coroner's Court.

One common complaint often heard from companies and organisations that have put together mountains of Health & Safety paperwork and documentation (usually without understanding what it all means) is, why should they bother – because nobody asks to see the paperwork?

The answer is: don't worry! The fact that nobody asks to see it is not the point.

The purpose of the paperwork is to act as your insurance after an accident, your evidence that safe systems of work were in place, and that you are doing things correctly!

16 THE HEALTH & SAFETY (INFORMATION FOR EMPLOYEES) REGULATIONS 1989

These regulations came into force in October 1989 and require employers to provide information to their employees with certain information relating to health, safety and welfare at work.

This information may be provided either by:

- Displaying the 'approved poster' or
- Providing the 'approved leaflet'.

Both of these are available from HMSO.

If you choose to comply with the regulations by displaying the poster, you must ensure that the poster is:

1) Kept displayed in a legible condition.
2) In a place which is reasonably accessible to all employees (e.g. the staff noticeboard or in the canteen).
3) In such a position as to be easily seen and read by all employees.
4) The current approved version (HSE may from time to time revise the contents of the the poster) was released on 1st October, 1999.

The poster should be completed with the address details of the local enforcing authority (HSE area office or local authority environmental health department) and the local Employment Medical Advisory Service (EMAS) office.

29 million man/days are lost each year through accidents at work.

17 THE EMPLOYERS' LIABILITY (COMPULSORY INSURANCE) ACT 1998

The act requires almost all employers to insure against liability for personal injury to employees.

The Act states that all employers, unless exempt from the provisions of the Act, must take out an 'approved' insurance policy with an 'authorised' insurer.

The insurer must provide the employer with a certificate of insurance that the employer must display at the work premises for the information of his employees.

The employer is also obliged to produce the certificate of insurance for inspection by an authorised inspector as and when required to do so.

Failure to comply with the terms of the Act is a criminal offence.

It is important to declare full details of the company's activities to an insurer or you may find your insurance is invalid. For instance, if you fail to mention that you carry out work at height in your insurance proposal and then try to make a claim after a climbing accident, insurance will only normally be available (at realistic terms) through a specialist music industry insurer who understands the risks.

Definitions

An *approved policy* is one which does not contain prohibited clauses, which would allow the insurer to disallow a claim under the policy.

An *authorised insurer* is a person or company lawfully carrying on an insurance business in Great Britain under Part 2 of the Companies Act 1967.

An *employee* is defined as: "an individual who has entered into or works under contract of service or apprenticeship with an employer .... whether such contract is expressed or implied, oral or in writing."

Exemptions

The following classes of employer are not covered by the provisions of the Act:

- Nationalised industries
- Local authorities
- Police authorities
- Businesses in Northern Ireland, the Isle of Man and the Channel Islands

The following types of workers are not covered by the provisions of the Act:

- People who are not employees as defined in the Act (e.g. sub-contractors).
- People employed in an activity which is not classed as a 'business' under the Act (e.g. domestic servants).
- People who are related to their employer as their husband, wife, father, mother, grandfather, grandmother, step-father, step-mother, son, daughter, grandson, granddaughter, step-son, step-daughter, brother, sister, half-brother, half-sister, etc.
- People who do not normally live in Great Britain and who are working here for fewer than 14 consecutive days.

Certificates of Insurance

The insurer is required to provide the employer with a certificate of insurance stating that the policy meets the requirements of the Act within 30 days of the commencement of the policy. The employer must:

- Display the certificate at the place of business for the information of his employees
- Produce on demand, the certificate or a copy for inspection by an authorised inspector.
- Copies of all old Employers' Liability Insurance certificates must be kept by the employer for 40 years!

Penalties

For *each day* that an employer is not insured under the terms of the Act he will be liable on summary conviction to a fine of up to £2,500.

Failure to produce a certificate of insurance for inspection by an authorised inspector and failure to display a current certificate of insurance each carry a maximum penalty of a £1,000 fine.

How do I know whether I am protected by Employers' Liability Insurance?

You will only be protected by your employer's liability insurance if you are an employee.

Sometimes it can be difficult to work out whether you are protected, especially if you generally think of yourself as self-employed. Whether or not you are covered by employers' liability insurance depends on your agreement with the person you work for. This agreement can be spoken, written or implied. It does not matter whether you generally call yourself an employee or self-employed and your tax status is irrelevant. What matters is the real nature of your relationship with the person you work for and the degree of control they have over the work you do.

There are no hard and fast rules about when you are an employee for the purposes of the employers' liability insurance. The following paragraphs may help give you some indication. However, if you have any doubts, you should seek legal advice.

In general, the person you work for may need employers' liability insurance to cover you if:

- They deduct national insurance and income tax from the money they pay you.
- They have the right to control where and when you work and how you do it.
- They provide most of the materials and equipment you need to do your job.
- They have a right to any profit you generate, although they may choose to share this with you through commission, performance pay or shares in the company. Similarly they will be responsible for any losses.
- You have to deliver the service yourself and cannot choose to employ a substitute.
- You are treated in the same way as other employees. For example, if you are doing the same kind of work under the same conditions as someone else who is an employee of the business.

In general, you may not be covered by employers' liability insurance if:

- You work for more than one customer and operate as an independent contractor. However, if you have more than one job you could still be an employee.

- You supply most of the equipment and materials you need to do the job.
- You receive the benefit from any profit you make rather than the person you work for, and you are personally liable for any losses.
- You can employ a substitute when you are unable to do the work yourself.
- The person you work for does not deduct income tax or national insurance. However, even if you are self-employed for tax purposes you may be classed as an employee for other reasons and your employer may still need employers' liability insurance to cover you.

You may not be covered by employers' liability insurance if you are a volunteer. Although the law may not require your employer to cover you if you are:

- A student working unpaid.
- Not employed, but taking part in a youth or adult training programme.
- A school student on a work experience programme.

In practice, many insurance companies will provide cover for such cases. If you are in one of those situations, you should ask the person you are working for whether their insurance covers you.

One difficult area is domestic help. If you work for more than one person, for example, if you are a cleaner or a gardener, you will probably not be protected by employers' liability insurance. However, if you work for only one person, they may be required to take out insurance to cover you.

Do I need Employers' Liability Insurance for all the people who work for me?

You are only required by law to have employers' liability insurance for people you employ. However, people who you normally think of as self-employed may be considered as your employees for the purposes of employers' liability insurance (but this can affect employment status for Tax purposes).

Whether or not you need employers' liability insurance for someone who works for you depends on the terms of your contract with them. This contract can be spoken, written or implied. It does not matter whether you usually call someone an employee or self-employed or what their tax status is. Whether you choose to call your contract a contract of employment or a contract for services is largely irrelevant. What matters is the real nature of your relationship with the people who work for you and the degree of control you have over the work they do.

In general, you may need employers' liability insurance for someone who works for you if:

- You deduct national insurance and income tax from the money you pay them.
- You have the right to control where and when they work and how they do it.
- You supply most materials and equipment.
- You have a right to any profit your workers make although you may choose to share this with them through commission, performance pay or shares in the company. Similarly, you will be responsible for any losses.
- You require that person only to deliver the service and they cannot employ a substitute if they are unable to do the work.
- They are treated in the same way as other employees, for example, if they do the same work under the same conditions as someone you employ.

In general, you may not need employers' liability insurance for people who work with you if:

- They do not work exclusively for you (for example, if they operate as an independent contractor).
- They supply most of the equipment and materials they need to do the job.
- They are clearly in business for personal benefit.
- They can employ a substitute when they are unable to do the work themselves.
- You do not deduct income tax or national insurance. However, even if someone is self-employed for tax purposes they may be classed as an employee for other reasons and you may still need employers' liability insurance to cover them.

In most cases you will not need employers' liability insurance for volunteers, although in general the law may not require you to have insurance for:

- Students who work for you unpaid.
- People who are not employed, but taking part in a youth or adult training programme.

- A school student on a work experience programme.

In certain cases they might be classed as your employees. In practice, many insurance companies will provide cover for people in these situations. If you have volunteers, students or non-employed trainees working for you, it is advisable to inform your insurance company and to consider carefully whether you should have insurance cover for them. You should also bear in mind that some of these people could be classed as your employees when you think about the amount of cover you need. Strangely enough, the above indicators are similar to the ones used by the Inland Revenue to assess if a person is employed or self-employed and to assess if a person is operating a business that should be operating under the IR 35 Tax regulations.

One difficult area is domestic help. In general, you will probably not need employers' liability insurance for people such as cleaners or gardeners if they work for more than one person. However, if you employ someone who works only for you, you may be required to take out insurance to protect them.

More recently, problems have arisen with the cost of insurance rising by huge amounts, increases of over 500% are not unusual.

There are four main reasons for these vast increases: the tragic events in New York City on 11th September 2001; an escalating number of large insurance claims; the proliferation in the number of companies advertising 'no win, no fee' legal representation to potential claimants; and the state of the finance markets (insurance companies play the stock market) – all have taken their toll.

The situation has now become so serious that some companies are unable to obtain insurance at realistic rates and in some cases are finding it impossible to even obtain a quote, thus forcing them to operate illegally without insurance. The Department of Works and Pensions have carried out an inquiry with a view to reviewing the whole situation regarding employers' and public liability insurance. The inquiry concluded that there had not been a general employers' liability market failure, that the overwhelming majority of firms can find insurance, albeit at a price, that levels of compliance remained high and that firms should 'shop around' for insurance.

In January 2004 the government announced proposed changes to the Employer Liability Insurance Act that could mean as many as 300,000 of the UK's smallest firms could be exempted from purchasing employers' liability insurance following a report issued by the Department of Works and Pensions.

The report proposes to exempt up to 300,000 limited companies that employ only the owner from the need to hold employers' liability insurance. At present all self-employed persons who operate under a Limited Company status are legally required to hold employers' liability insurance (as they are employed by the Limited Company). Public liability insurance on its own is *not* sufficient.

Historically, employers' liability insurance has fulfilled three functions:

a) Securing compensation for employees injured or made ill at work
b) Spreading the risk for employers
c) Making the polluter pay

A fourth function would make an effective tool to solve problems and improve Health & Safety.

As part of the Government's review of the troubled employers' liability insurance industry the following objectives have been set:

- To help us through the transition – short term help for business and industry.
- To introduce more risk-based underwriting, forging a more effective link with safety; and
- To stabilise inflationary pressures – legal costs, compensation and long tail occupational disease risk.

All to create a better linkage between Health & Safety performance and the cost of insurance!

Health & Safety is a business risk, but effective Health & Safety management is a business benefit.

The report concludes that in order to keep premium increases under control savings must be made by:

- Tacking the legal and administrative costs associated with employers' liability insurance.
- Providing employers and insurers with more information an good Health & Safety management practices, so that good Health & Safety performance can be reflected in the price of premiums; and
- Improving the Health & Safety performance of UK businesses to reduce the number of injuries and cases of ill health that could give rise to claims.

Safety Minister, Des Brown, said that the government recognises that too many businesses have faced steep price increases; late renewals and premiums that fail to reflect their Health & Safety record.

The report also mentions details of 'Making the Market Work', an initiative launched by the Association of British Insurers that enables trade associations and similar bodies to have Health & Safety schemes assessed by an ABI committee against best practice features that insurers expect to see in place. This initiative has been introduced in an attempt to lower insurance costs for those companies and organisations with effective Health & Safety management systems in place.

Insurance companies are now advising self-employed operators (such as stage, tour or production managers) who instruct, control or supervise other crew to take out employers' liability insurance (in addition to public liability insurance) because they are taking on responsibility for the way the crew operate. If one of the crew they are supervising, controlling or instructing, suffers an injury, it is then possible for them to sue the self-employed (freelancer) who has been supervising or instructing them for damages.

Self-employed operators need to have their own public liability policy to protect themselves in the event of an injury or damage to a third party or their property. If the self-employed person does not have the required insurance, they are taking a high risk and may find it difficult to obtain work, as no one in their right mind will engage a contractor who is not properly insured, even though it is not a legal requirement for self-employed contractors to hold public liability insurance. Some companies who use the services of self-employed contractors are also insisting that the self-employed contractor insures any equipment they use that belongs to the company that has hired their services.

It may be possible for the self-employed contractor to include a clause in his or her contract (to supply goods and services) that states they will not be insured against damage to a company's equipment while they are in control of that equipment and that they will not hold employers' liability insurance to cover crew members but this remains a grey area and it is not certain if such clauses will offer the required legal protection.

For further information about insurance, contact your insurance company, broker, agent, or the Association of British Insurers.

18 THE CONSULTATION WITH EMPLOYEES' REGULATIONS 1996

This regulation came into force on 1st October 1996 and marks the completion of the government's implementation of the European Community Framework Directive 89/391/ECC on consultation rights for employees. The 1977 Safety Representatives and Safety Committees' Regulations still remain in place but it is expected that this regulation will be merged with the Consultation with Employees' Regulations 1996 in the not too distant future to give one set of new regulations, but at the time of writing, this plan has been temporarily shelved.

Consultation with those doing the job is crucial to avoid accidents and ill health, and pooling knowledge and experience is a key aspect of risk control. From an *employee's* point of view, a safe operation means a more secure work environment due to the knowledge that safety concerns will be met with the appropriate actions. From the *employer's* point of view, a safe operation means a well-motivated workforce, fewer costly accidents/incidents, more efficiency and greater productivity. Consultation not only benefits the employer but also the customers, as it means that not only are customers' requirements being met, but also some of the customers' legal obligations will have been fulfilled.

The regulation requires all employers to consult with employees who are not covered by representatives appointed by a recognised trade union on matters relating to health and safety at work.

It gives the employers the choice to consult directly or via *representatives elected by the employees* they are to represent. This means that employers must provide the information necessary for employees to participate fully and effectively in consultation on health and safety matters and provide employee representatives with training, time off and the facilities to enable them to carry out their duties. All this must be at the employer's expense and where possible in normal working hours.

For a small organisation such as most of those we find within our industry it is quite easy for the employer to consult directly on a one-to-one basis.

Employers need to consult with employees in the following areas:

- Information about any measure at the work place which may

substantially affect employees' health and safety.

- Information about plans for appointing a competent person to help the organisation comply with Health and Safety requirements.
- Information on the risks to health and safety and the preventative measures.
- Information on the planning and organising of Health and Safety training. This should include training for present *and developing* Health & Safety legislation.

Employee representatives must be given suitable training to give them the adequate understanding of Health & Safety legislation they require and must have to enable them to do their jobs, i.e. a NEBOSH or British Safety Council certificate, and the employer must pay for this training that should be carried out during normal work hours if possible. The employee representative must be paid at the normal rate during such training.

Information on the Health and Safety consequences for employees of new technology that the organisation plans to bring into the workplace

The employer is required to provide the information necessary to enable staff or their representatives to participate fully in the consultation. This should include reportable accidents and diseases (see RIDDOR) and general material on hazards and risks, but not any information which is personal to an individual without that person's permission.

Representatives have two key functions: making representations to the employer on hazards which could affect the staff represented and general health and safety matters, including any on which the employer initiates consultation and also represent staff in discussion with appointed inspectors (typically Environmental Health Officers).

All consultation meetings should be recorded in writing, remembering that provision of information is a one-way process. Consultation, on the other hand, is a two-way process that involves listening to employees' views and considering what employees say before reaching a decision. Management must listen, evaluate and where appropriate act upon issues raised by employees.

Consultation does not remove the requirement for employers to provide information and training to employees as required under the Management of the Health and Safety at Work Regulations.

The regulations state that employers must consult 'in good time' and provide information about what they propose to do and allow employees or their representatives time to give their views about the matter in the light of that information *before* the employer acts.

These new regulations do not apply to the self-employed. However, case law has been established that workers who are categorised as self-employed for tax or other purposes may be employees in respect of Health & Safety law. It can depend on individual circumstances such as the workers' degree of independence, and if workers are providing and using their own tools, etc. It may be a good policy to include the self-employed, especially if they have been engaged to provide services by the employer for a long period or are exposed to particular risks and hazards at work.

Training, as will become obvious as you read on, is one of the main ways of controlling risks, and employers have a legal duty to train staff and provide them with information on the risks they face at work and how to avoid or reduce these risks.

As mentioned, all training must be given to staff during normal working hours and staff must be paid at the normal rate during training.

Employers cannot expect staff to pay for their own training, but employment contracts may require staff to repay the cost of any training courses if the employee leaves the company or business before a given time period, say, for example, two years.

The guidance to the regulations explains how *employees* should go about electing a representative and what training and facilities an employer is obliged to provide under the regulations for these representatives.

I would urge all *employers* to become conversant with these regulations before you land yourself in serious trouble. They *do* affect YOU!

If a new set of regulations is introduced after merging these regulations with the Safety Representatives and Safety Committees' Regulations 1997, then it is expected they would give employees in a workplace where a trade union is not recognised the right to choose how, and by whom, they should be represented on Health & Safety issues. In addition, the new regulations are expected to expand the powers of trade union appointed safety representatives to those elected by employers. This would mean that in future *all* safety representatives would have the right to:

- Receive information from Health & Safety inspectors
- Represent employees in meetings with enforcing authorities

- Attend meetings of workplace safety committees

If adopted, the proposals in the new regulations would also give safety representatives (from a staging, lighting or audio company for instance) the power to inspect the workplaces that are owned or in control of a third party (i.e. a venue or festival site), but where the employees that they represent are engaged!

The HSC is convinced that safety representatives have a major role to play and will have a serious influence on future Health & Safety strategies.

A workforce that is actively involved in Health & Safety decision-making is more likely to be fully cooperative in implementing and maintaining Health & Safety management systems.

19 SAFETY REPRESENTATIVE AND COMMITTEE REGULATIONS 1977

Until the formation of the Roadies' Union (Road-crew Provident Syndicate) we didn't have much contact with trade unions in our industry. I'm not sure if that's a good or bad thing, but what I do know is that recognised trade unions may appoint safety representatives. This does not prevent employers, where there are no recognised trade unions, from appointing safety representatives from their workforce. It was expected that the 1977 Safety Representatives and Safety Committees Regulations would be merged with the Consultation with Employees Regulations 1996 in the not too distant future to form one set of new regulations, but at the time of writing the plan to do this has been temporarily shelved by the HSE.

The main difference between the two types of safety representative is that the former has legal functions (but no duties), whereas the latter type has the role assigned to them by their employer.

The number of safety representatives will depend on:

- The size of the workplace
- The size of the workforce
- The hazard potential

The functions of Safety Representatives are:

- To investigate potential and dangerous occurrences.
- To investigate health, safety and welfare complaints from employees.
- To make representations to employers (on behalf of employees)
- To be kept up-to-date with information resulting from visits by HSE inspectors.
- To be consulted about the introduction of Health & Safety measures, nomination of 'competent persons', planning and organisation of training, and the introduction of new technology.

In addition, safety reps. are entitled to carry out inspections of the workplace(s), often advising, and in consultation with, the employer. In general, this will not be more frequent then three monthly intervals.

To enable safety representatives to carry out their duties they are entitled to see (and copy) relevant Health & Safety documents other than:

- information that would be against the interest of national security or the employer's undertaking.
- information, the disclosure of which would be unlawful.
- personal information
- information obtained for taking or defending legal proceedings.

Safety Committees are established by either a recognised minimum of two trade union representatives asking for a committee to be formed (to be done within three months of the request), or at the instigation of the employer.

In either case, a committee will only function as well as its organisation allows. Its composition, agenda, the frequency of meetings, keeping and distribution of minutes, etc. must be clearly defined at an early stage.

The number of committees will depend on the size of the organisation and/or the number of sites.

Chairmanship of the committee must be such that its consideration decision and recommendations have authority and can't be ignored.

The relationship of the safety manager/adviser, doctor and/or nurse should be clearly established as ex-officio members and advisory only.

Equality of representation between safety representatives and others should be carefully considered.

Union-appointed safety representatives will receive training from their union (when their employer must be given time off with pay).

Other safety representatives should be adequately trained for their role and means of carrying it out. This training is in addition to any other general or specific training needs.

20 HEALTH & SAFETY (FIRST AID) REGULATIONS 1981

This is one of the simpler sets of regulations under the umbrella of the HASAWA that both companies and the self-employed should have no problem complying with. First aid facilities form part of a Health & Safety management system but should never be considered as a substitute to effective Health & Safety management, as prevention is always better than cure. Possessing first aid skills and basic equipment can save lives not only at work but also in our home and social lives. It is hoped that this will encourage people to obtain the required qualifications and for employers to make their companies and premises compliant with the regulations.

This set of regulations came into operation in 1982 and placed various obligations on employers relating to the provision of first aid in the workplace. These regulations apply to almost all workplaces; event organisers and promoters should refer to the Event Safety Guide for details of the first aid and medical facilities required for members of the public attending an event.

Definition

'First Aid' is defined as:

a) in cases where a person will need help from a medical practitioner or nurse, treatment for the purpose of preserving life and minimising the consequences of injury and illness until such help is obtained, and

b) treatment of minor injuries which would otherwise receive no treatment or which do not need treatment by a medical practitioner or nurse.

Under these regulations, there is a legal requirement for employers (promoters, artists, managers, venues or service companies) to provide or ensure that there are provided, such equipment and facilities as are adequate and appropriate in the circumstances for enabling first aid to be rendered to their employees if they are injured or become ill at work. This applies even when they are working away from the employers main place of business such as on site or on a tour. They also have a responsibility to inform their employees of the arrangements that have been made in connection with the provision of first-aid, including the location of equipment, facilities and the identity and location of personnel trained to administer first aid. This should be part of the induction training given to new employees when they first join the company or business.

Employers also have a legal duty to provide, or ensure that there is provided, such numbers of suitable persons as are adequate and appropriate in the circumstances for rendering first aid to their employees if they are injured or become ill at work. For this purpose, a person shall not be suitable unless he/she has undergone:

a) Such training and has such qualifications as the Health & Safety Executive may approve for the time being in respect of that case or class of case, and
b) Such additional training, if any, as may be appropriate in the circumstances of that case.

This in other words means 'qualified' First Aiders or 'Appointed Persons'.

Appointed Persons may be sufficient in low risk situations, when the work place is equipped with a telephone to contact the emergency services and access is not a problem.

Qualifications can be obtained on courses run by the St. John Ambulance Brigade, the British Red Cross, St. Andrews Ambulance Association, as well as many colleges and commercial companies authorised by the HSE. The qualification is the full First Aid at Work certificate or the Appointed Person certificate (as approved by the Health & Safety Executive).

Training involves an intensive four-day course with an exam at the end for the full First Aid certificate. The costs are currently in the region of £150-£200 per person. The certificate is valid for three years, after which a two-day course (with an exam) is required to renew the certificate. The revised Approved Code of Practice recommends that a refresher course is taken every 18 months, this is particularly important for those who get little or no practical experience between exams. Not bad advice.

The Appointed Person course is only 4-8 hours in duration and costs £25-£40. These courses are well worth doing and represent excellent value for money.

We should have as many qualified first aiders as possible within the industry. Employers are required to pay course costs for their employees and some even pay them a little extra as first aiders in addition to their normal wages. I wonder when we will see good practice like this within our industry?

During most events we end up working at there is usually adequate first aid provision during the event itself, but the danger times are during load in/outs before the medical teams and stewards arrive or after they have left.

An employer has to provide the *minimum* of at least *one Appointed Person* when employees are at work. To cover ourselves in a touring/festival situation where employees of more than one employer (and self-employed persons) are working together, and to avoid the duplication of provision, then an arrangement should be made whereby one of the employers or the 'tour/festival/show production' provides all the necessary First Aid equipment and facilities. This agreement should be in writing and a copy kept by each employer (and self-employed person) concerned. Where such an agreement is made, each employer should inform his/her own employees of the arrangements for first aid.

Category of risk	Numbers employed at any location	Number of First Aid personnel
Lower risk e.g. shops, offices, libraries, etc.	Fewer than 50	At least one appointed person.
	More than 100	One additional First Aider for every 100 employed.
Medium risk e.g. light engineering and assembly work, food processing, warehousing	Fewer than 20	At least one appointed person.
	20-100	At least one First Aider for every 50 employed (or part thereof).
	More than 100	One additional First Aider for every 100 employed.
Higher risk. e.g. most construction, slaughterhouse, chemical manufacture, extensive work with dangerous machinery or sharp instruments.	Fewer than 5	At least one appointed person.
	5-50	At least one First Aider.

The Approved Code of Practice for the first aid regulations gives numbers of First Aid personnel to be available at all times people are at work, based on assessments of risk and numbers of workers, shown in the above table.

Where there are special circumstances, such as remoteness from emergency medical services, shift work, or sites with several separate buildings or work areas, there may need to be more first aid personnel than set out below. Increased provision will be necessary to cover absences.

Companies should carry out Risk Assessments to establish what level of first aid cover is required; the above table is only intended to offer guidance to the minimum levels of first aid cover and is not intended as a replacement to a thorough Risk Assessment.

Event organisers and promoters should remember that just because they have stewards who may be trained Appointed Persons or first aiders does not mean they can cut down on the number of first aiders or Appointed Persons at the event.

Casualties with minor injuries of a sort they would attend to themselves if at home may wash their hands and apply a small sterilised dressing from the first aid box. A record must still be kept in the accident record book and any use of first aid equipment reported to the person responsible so that it can be replaced as soon as possible.

The guidance to the regulations states that First Aid kits should contain only suitable equipment that the First Aider has been trained to use. Suitable equipment will consist of the following items *and nothing else*:

- A guidance card
- Individually-wrapped sterile adhesive dressings (assorted sizes) appropriate to the work environment (which may be detectable for the catering industry)
- Sterile eye pads (with attachment)
- Individually wrapped triangular bandages
- Safety pins
- Sterile individually wrapped un-medicated wound dressings (small, medium and large)
- Disposable latex gloves.
- Blunt ended stainless steel scissors (minimum length 12.70cm)
- Plastic disposable bags for soiled or used first-aid dressings.
- Where no mains tap water is available for eye irrigation, sterile water

or sterile saline (0.9%) in sealed disposable 300ml bottles (at least 900ml should be provided).

And for Travelling Kits only (in addition to the above):

Individually-wrapped moist cleansing wipes (not impregnated with alcohol).

Under no circumstances should antiseptics, aspirin, painkillers, eye baths/cups, creams, tablets, sprays or other medications be kept in first aid kits.

It is better to have a number of small first aid kits distributed around the workplace as opposed to one large one, and don't forget to equip all vehicles with a first aid kit.

The treatment of minor illnesses and the administration of tablets and medication falls outside of the definition of the Health & Safety (First Aid) Regulations 1981 and forms no part of the training for a first aider. Often tour caterers or production provide items such as painkillers or preparations for colds or 'flu etc. to tour personnel, but this is a dangerous practice and is not recommended.

In addition to the equipment listed above, soap, water and disposable drying materials should be provided for first aid purposes.

A First Aider should be in charge of the first aid box, and a system (recorded) must be in place so that any items used must be replenished as soon as possible to ensure that there is always an adequate supply. First aid kits should be positioned in clearly identified and readily accessible locations convenient to the majority of the work force or where there is the greatest risk of injury occurring.

First Aid kits, rooms and facilities should be clearly marked in accordance with the Health & Safety (Safety Signs & Signals) regulations; the marking should be a white cross on a green background.

Employers should keep records of First Aiders' certification and training as well as displaying a notice listing first aiders (with contact details such as extension numbers) in the workplace.

Self-employed persons have a legal responsibility under the Act to provide, or ensure that there is provided, such equipment, as is adequate and appropriate in the circumstances to enable him/her to render first aid to him/herself while he/she is at work. As mentioned before, any agreement about the provision of fist aid facilities between self-employed and those using the services of the self-employed should be in writing. If no agreement exists, then the self-employed person must provide his or her own first aid facilities.

For further information, please refer to the Guidance and Approved Code of Practice to the Health & Safety (First Aid) Regulations 1981 published by the HSE.

A frequently asked question is: "Can I be sued or prosecuted for giving first aid incorrectly?" The answer is that if you are qualified (under the Regulations) you are covered by employers'/public liability insurance while at work and it must be proven that you acted with malicious intent. A court of law would normally consider you acted under the Good Samaritan principle unless proven otherwise.

21 RIDDOR 95 REPORTING OF ACCIDENTS

Under the 1975 Social Security Act, employers have a duty to make a record of *all* accidents and in some cases these need to be reported to the local authority or the HSE under the Reporting of Injuries, Diseases, and Dangerous Occurrences Regulations 1995. The regulations make the reporting of certain instances of injury, disease and dangerous occurrence a legal requirement. This information is used by the authorities to identify trends in work related accidents and ill health so that they can help and advise companies and individuals on appropriate preventative measures.

The chief purpose of reporting to the HSE is so that they can identify how problems arise and show up any trends. This then helps them to perform their various accident prevention activities. Keeping an accident book will also help employers identify trends and show problems as well as acting as further evidence as to the 'quality' of the business.

The regulations place duties on employers, the self-employed and people in charge of work premises. Basically, they require that if a 'reportable' accident, injury, dangerous occurrence or case of disease happens in your workplace, you inform the 'enforcing authority' and keep a record of all the details.

The 'enforcing authority' will be either your local HSE area office or the environmental health department of your local authority depending on the nature of your business. Broadly speaking if your business fits into one of the following categories then the report should be made to the environmental health department of your local authority:

Hotel and catering
Office based
Places of worship
Residential accommodation (does not include nursing homes)
Retail or wholesale
Sports and leisure (including gigs, shows, concerts and festivals)
Warehousing

If your business does not fall into one of these categories or if you are unsure who the relevant enforcing authority is, then send the report to your local HSE office.

RIDDOR 95 places the responsibility for making the report with the 'responsible person'. In most instances the responsible person will be the employer, manager, or person in charge of operations at the site where the event occurred. Failing to make a report under RIDDOR, not having a competent person appointed to make reports, or not having a reporting procedure in place can result in a fine of £5000.

Well-kept records of accidents and ill health can play a useful part in the risk assessment process from the employer's point of view.

The recording of an accident in the accident report book may strengthen any later insurance claim or even prove if the accident actually took place in work time. For example, a major opera company used to use high quality marine plywood with which to construct its stage sets, but to save a few pounds they changed to using a poorer quality, shuttering ply (that is well known for splintering). One day, a carpenter employed by an opera company received a splinter in his finger after handling the wooden stage sets. He quickly removed the splinter and thought nothing more of the incident until a few days later when his finger turned septic and started to swell up to the size of a football causing him considerable pain. After examination by a doctor, the carpenter was taken to hospital and his finger was amputated, such was the severity of his condition. Unfortunately, the poor chap had failed to fill in the accident report book at work and his employer's solicitors claimed he may not have received the splinter at work and, therefore, he was unable to claim any compensation for an industrial injury.

It is the legal duty of an employee to inform his employer if he is injured at work. This can be done by filling in the accident report book, but anyone can do this on his/her behalf. Accident report books can be obtained from HMSO Publications (Form B1 510). The HSE introduced a new style Accident Report Book in 2003 and this new style book must now be used. The reason for the change was to meet the requirements of the Data Protection Act 1988 that forbids the recording of personal data such as a workers' addresses in the accident book. The new-style book can be obtained from HSE Books. (ISBN 0-7176 2603-2) and has perforated sheets to allow employers to remove pages in order to comply with the law covering the Data Protection Act.

The accident report book should be kept where any employee can obtain access to it, such as with the first aid kit. It can be filled in by anyone, not just an accident victim or a first aider, but if the person filling it in did not see the accident happen he or she should write in the third tense, for example "Fred *said* that he hurt his leg when he fell off the ladder". When the book is full it should be kept on file by the employer for three years after the date of the last entry.

As mentioned, certain injuries, diseases or dangerous occurrences *must* be reported under RIDDOR to the Health & Safety executive. Reports should first be made by the quickest means possible to the Environmental Health department of the local authority (see phone book) and then a report made on form F2508 for accidents or dangerous occurrences or F258A for cases of disease; these are obtainable from HSE Books. A permanent record must be kept of all reportable injuries, ill health and dangerous occurrences and the records kept by a responsible person as defined by the regulations, usually the employer. The HSE has now started an even simpler method of reporting with the introduction of an Incident Contact Centre. The contact details are: tel: 0845 300 9923 fax: 0845 300 9924 email: riddor@natbrit and a website at www.riddor.gov.uk

The person making the report should ensure that records of all the events are kept on file. Records should detail:

- The date and method of reporting
- The date, time and location of the incident
- Certain personal details of those involved
- A description of the event/disease

The regulations do not stipulate the form in which records should be kept. Saving a copy of the file you send to the enforcing authority will be sufficient to meet this requirement.

There are many types of injuries, dangerous occurrences and disease that must be reported but (other than injury) most of them are unlikely to affect our industry. A full list can be found in the HSE publication *A Guide to the Reporting of Injuries, Diseases and Dangerous Occurrences Regulations 1995.* The regulations changed again in 2003 and it is now required to report acts of psychological violence. The following must be reported:

- Any fatality.
- Any fracture other than to fingers, thumb or toes.
- Any amputation
- The loss of sight (temporary or permanent)

- A penetrating injury to the eye, or a chemical or hot metal burn to the eye.
- An injury resulting from electric shock or electrical burn leading to unconsciousness, or requiring resuscitation or admittance to hospital for more than 24 hours.
- Loss of consciousness caused by asphyxia or by exposure to a harmful substance or biological agent.
- Acute illness requiring medical treatment or loss of consciousness that has resulted from the absorption of any substance by inhalation, ingestion or through the skin.
- Acute illness requiring medical treatment where there is a reason to believe that this resulted from exposure to a biological agent or its toxins or infected material.
- Any other injury that results in the person injured being admitted immediately into hospital for more than 24 hours.
- An injury (other than one of those listed above) that results in incapacity from work for more than three days.
- The collapse of, the overturning of, or the failure of load bearing parts of lifts and lifting equipment.
- Explosion, collapse or bursting of any closed vessel or associated pipe work.
- Failure of any freight container in any of its load bearing parts.
- Plant or equipment coming into contact with overhead power lines.
- Any unintentional explosion, misfire, failure of demolition to cause the intended collapse, projection of material beyond a site boundary, injury caused by explosion.
- Failure of industrial radiography or irradiation equipment to de-energise or return to its safe position after the intended exposure period.
- Collapse of a scaffold over five metres high, or erected near water where there could be a risk of drowning after a fall.
- Unintended collision of a train with any vehicle.
- A dangerous substance being conveyed by road is involved in a fire or is released.
- Explosion or fire causing suspension of normal work for over 24 hours.
- Sudden, uncontrolled release in a building of 100kg or more of flammable liquid; 10kg of flammable liquid above its boiling point;

10kg or more of flammable gas; or of 500kg of these substances if the release is in the open air.

Reportable diseases include:

- Certain poisonings
- Some skin diseases such as occupational dermatitis, skin cancer, chrome ulcer, oil folliculitis/acne.
- Lung diseases including asthma, farmer's lung, pneumoconiosis, asbestosis, mesothelioma.
- Infections such as: leptospirosis; hepatitis; tuberculosis; anthrax legionellosis and tetanus.
- Other conditions such as: occupational cancer; certain musculoskeletal disorders; decompression illness and hand-arm vibration syndrome.

The self-employed must report accidents to the company that has contracted their services and make reports under RIDDOR, if required. Any reportable injury that has occurred outside of the UK should be reported as soon as possible upon return.

People seem to be frightened to make these reports as they think they with get into trouble with the HSE or that the HSE will be down on them like a ton of bricks and carrying out investigations. Well, the truth is that it is very unlikely that the HSE will want to investigate your business unless there has been a death or a very serious accident. They are more interested in gathering accident details and information so that they can use this information in their accident prevention schemes and to do this they need to look at the 'big picture' that helps to identify trends and patterns. Even the simple accident book kept at your premises or on site can be used in much the same manner.

Let's think for a minute or two about asbestos. No doubt even the mention of the word brings thoughts of cancer and death to your mind but there was a time when asbestos was the new wonder material that could be moulded, bent, shaped, was strong yet could also be soft and flexible and heat resistant, it was even used to build whole schools!

A short time later people all over the country started getting strange illnesses and cancer, and it was then that a common link was found between all these different cases. The link was that they all used and had contact with asbestos, and had it not been for the work of the HSE gathering all this accident information, it could have been much longer before the discovery was made and action to prevent asbestos contamination brought into place.

After any serious accident the site may need to be preserved like the scene of a crime so that a full investigation can take place. The HSE has the power of entry and can enlist the police to help them if required. They may take photos or take away any items they think necessary to help with an investigation.

It is our legal duty to cooperate and assist HSE officers during their work; obstruction is a breach of the law.

Health and Safety Executive officers and Environmental Health officers have the same powers of enforcement but it is the local authority officers (EHOs) who enforce entertainment licence conditions *as well as* Health & Safety at the events we work at. HSE officers are much more serious and not to be messed with, but you only tend to see them after an accident or if there is a major problem.

Stress is now the biggest cause of absenteeism from work. Risk assessments must be made for stress and action taken by employers to remove or reduce stress at work.

22 ACCIDENT INVESTIGATION

All accidents, regardless of size and severity need to be thoroughly investigated so that lessons can be learnt and similar accidents prevented. To do this, a simple accident investigation strategy needs to be followed.

If someone suffers an accident on your premises there can be a lot of initial confusion and disruption, with a number of people attending the scene in a short space of time. It is therefore important to put together an action plan and convey this to employees to ensure that should an accident occur the company remains in control.

The HSE had proposed new regulations that would have made it the responsibility of all companies with 40 or more employees to carry out accident investigations for any accidents that took place. These new regulations have now been postponed but may be introduced at a later date. Had the new regulations been introduced, it would have required an investigation to be carried out by a competent person and by any required experts that may have been required to be brought in to assist.

Companies with fewer than 40 employees would also have been required to investigate accidents up to (but not including) major accidents. A major accident is defined in RIDDOR as an accident that involves the victim being unable to carry out his or her *normal* work for three or more days.

Immediate Response

Immediately after an accident the first response must be to make the area safe, attend to the injured, call for assistance (ambulance) and preserve any evidence.

Try not to move anything unless absolutely essential for safety and close off the area; hazard warning tape can be used for this purpose.

Gather Information

In addition to any investigation carried out by the enforcing authority, you should always carry out your own internal investigation. The next step in this process is to gather information which can be gained from:

- The people involved

- The place where the accident happened, from the
- Plant, equipment and materials involved, and from existing
- Procedures and paperwork.

First the people involved must be formally interviewed as soon as possible; the purpose of the interview is not to apportion blame but to find out the real facts. A written signed and dated statement should be taken from all involved.

It is best to carry out the interview on site before memory fades. Witnesses should be kept apart to prevent distortion from rumour, speculation and collaboration, obviously any casualties may have to be interviewed when they are fit enough and/or after they have received medical treatment.

The place where the accident took place should be examined in detail before anything is moved. Drawings, sketches, measurements, notes and photographs can all help determine the cause and provide ideas to prevent further accidents. Similarly, any plant, equipment or materials should also be examined in the same way for damage, clues and other information.

Finally, any paperwork such as Health & Safety policies, company or local rules, risk assessments, training records, method statements, records of test, inspection and examination, qualifications, permits to work etc. all provide information and clues as to what has gone wrong or which systems (if any) have failed.

All the gathered information should be stored securely, since it may be required by the authorities, the HSE or further accident investigators, particularly if prosecutions or insurance claims are likely.

Analysing The Information

By studying all the assembled information it should be possible to gain an insight into what happened as the events unfolded. This will not only show the main cause of the accident, but will also point to what remedial action needs to be taken to prevent similar accidents.

Identify The Causes

Having analysed the information it should be possible to identify not only the main cause of an accident but also the underlying causes, such as a lack of safety culture, training, local rules, other safety control systems and even simple operator error due to lack of concentration.

Make Recommendations

The next step is to make recommendations to senior management about what needs to be done to prevent further accidents. This may include a review of 'local rules', the introduction of further training, safety signs and other safety control systems. The recommendations should be simple, clear and concise.

Carry Out Effective Action

The final step is the implementation of the recommendations, the recommendations need to be prioritised and the implementation timetabled and monitored to ensure that it is effective.

HSE Investigations

The HSE can and do carry out accident investigations, but usually only after more serious accidents. As part of an accident investigation the HSE can examine the books and financial records of a business to establish exactly how much the business was spending on Health & Safety. There is no set figure, but the amount should reflect the size of business, number of employees and the level of risk involved with the business activities. It should be possible to ascertain how much is being spent on training, safety consultants, personal protective equipment and other Health & Safety matters from the company's books and financial records. The latter could be used to determine how much the company had invested in Health & Safety.

If an accident is serious enough for the enforcement authorities to be involved and they intend to carry out an investigation, you should:

Appoint a member of senior management as a spokesperson to deal with all requests for information, or interviews by the enforcing authority and the press releases.

- Contact your safety adviser, solicitors and insurers straightaway.
- Advise your managers that enforcing authorities do have statutory powers to enter premises, seize documentation and speak to employees.
- Update your staff and warn them they may be asked to give an interview. Tell them they can submit a written response or request more time before signing the declaration of truth.
- If documents or materials are seized, ask that only copies are taken or take copies yourself. Make a note of all documentation taken and by whom..
- Set up a separate file to maintain all records reviewed by the enforcing authority and copy interviews. Include copies of relevant training

records, Health & Safety procedures, policies and risk assessments. Remember that civil claims can be brought a long time after the incident.

- Obtain written conformation from the enforcement authority before changing or repairing any equipment or machinery. This is best done by your solicitors to maximise the chance of keeping information confidential.

23 THE WEATHER, ENVIRONMENTAL AND BIOLOGICAL HAZARDS

When planning festivals or outdoor events, the weather, just like any other hazard, needs to be taken into consideration and risk assessments and contingency plans prepared. Your risk assessment can show that you have obtained and studied meteorological office reports and have prepared accordingly with items such as temporary roadway and hard standing for site vehicles as well as checking structural reports of stages and structures for wind or even snow loadings, etc. Staff working outdoors should also be made fully aware of possible adverse weather conditions and how to deal with them.

When doing your Risk Assessment for outdoor shows, consider stewards working in the pit. This area can become treacherous with wet and mud if not decked, boarded or covered with sand or possibly straw (provided of course the fire hazard of using straw is considered and assessed).

Clearly there is little that can be done in terms of weather prevention, but production managers must ensure that adequate waterproof covering is provided on stages and employers must provide a place for workers to rest away from weather hazards.

Driving conditions are more dangerous, particularly on green field sites where rain and mud can make them very slippery and the danger of overturning on a slope becomes very real. The use of a temporary roadway should be considered in the Risk Assessment, remembering that at least one fatal accident has occurred when trying to pull trucks out of the mud with the aid of a forklift truck!

It is obvious that rain and electricity do not mix. For added protection, all hand tools should be of the 110v type and protected by an RCD.

People tend to cut corners when it's cold and wet to get the job done more quickly, but at no time must an employer or production manager allow Health & Safety to be compromised when the weather gets bad. Remember, in cold weather the wind can reduce temperatures even more. This effect, known as the wind chill factor, must be treated seriously. The right clothing can make a job much more bearable and the chances are that staff will function in a safer manner as well as being more productive. A full set of waterproofs – jacket with hood, over-trousers and footwear such as Wellington boots, (even

Equivalent Temperature (°F)

Wind Speed MPH																	
Calm	35	30	25	20	15	10	5	0	-5	-10	-15	-20	-25	-30	-35	-40	-45
5	32	27	22	16	11	6	0	-5	-10	-15	-21	-26	-31	-36	-42	-47	-52
10	22	16	10	3	-3	-9	-15	-22	-27	-34	-40	-46	-52	-58	-64	-71	-77
15	15	9	2	-5	-11	-18	-25	-31	-38	-45	-51	-58	-65	-72	-78	-85	-92
20	12	4	-3	-10	-17	-24	-31	-39	-46	-53	-60	-67	-74	-81	-88	-95	-103
25	8	1	-7	-15	-22	-29	-36	-44	-51	-59	-66	-74	-81	-88	-96	-103	-110
30	6	-2	-10	-18	-25	-33	-41	-49	-56	-64	-71	-79	-86	-93	-101	-109	-116
35	4	-4	-12	-20	-27	-35	-43	-52	-58	-67	-74	-82	-89	-97	-105	-113	-120
40	3	-5	-13	-21	-29	-37	-45	-53	-60	-69	-76	-84	-92	-100	-107	-115	-123
45	2	-6	-14	-22	-30	-38	-46	-54	-62	-70	-78	-85	-93	-102	-109	-117	-125

COLD / VERY COLD / BITTER COLD / EXTREME COLD

WIND CHILL CHART

Wellingtons are available with steel toe caps) is required in wet weather. Underclothes may need to include thermal underwear and a hat and gloves may be required depending on conditions. These are all classed as items of PPE.

Working in the sun is a lot nicer but has its own problems: it's very easy to get sunburnt or even sunstroke and thousands of people get skin cancer every year.

Employers must advise workers that long term exposure to the sun may cause sunburn, blistering, skin aging, and in the long term, exposure can lead to skin cancer. There are 40,000 new cases of skin cancer diagnosed each year, making it the most common type of cancer. Workers should keep themselves covered up including the head, neck, arms and legs, they should use a sun block cream (yet more PPE) of at least factor 15 and drink plenty of liquid (but not alcohol) to prevent dehydration and stay in the shade whenever they can. If possible, plan operations so that is not necessary to be outside in the sun during the middle of the day. I know of one stewarding company that issues its staff with polo shirts with collars that can be turned up to protect the neck from the sun. Good idea!

There are people who worked on building sites years ago who have now developed skin cancer due to over exposure to the sun while working. These same people are now suing past employers for not warning them of this danger. Crew are advised to examine their skin regularly for abnormalities – for example moles or unusual spots that change size and, if necessary, seek medical attention.

Working in hot weather conditions can be very stressful and judgement can be badly impaired with the onset of even minor heat exhaustion and sunstroke with its 'flu-like symptoms.

Exposure to bright sunlight can also cause damage to the eyes. Some of this damage is short term, but some may cause derogation and irreversible long-term damage to the eyes and vision.

Hats with peaks can do a lot to shade the eyes from the damaging rays that cause the damage, as can quality sunglasses.

Many accidents involving forklift trucks have been caused by the driver being dazzled by the sun; sunglasses could have prevented an accident.

When choosing sunglasses, select a well made, well-fitting pair that have a 'wrap around' feature to the lenses and that conform the British or EN Standards.

After long dry periods, an outdoor site may become like a desert and dust may start to blow around and become a discomfort for workers and the public alike.

Consideration should be given to send a water tanker with the ability to spray water to keep down the dust around green field sites. Drivers of open vehicles such as tractors and staff working in car parks on traffic management duties will need dust masks in such conditions.

High wind can be an extreme danger to outdoor stages, tents and structures. One of the control systems can be the removal of waterproofing sheets to allow the wind to blow through. For anyone climbing an outdoor structure, the rain, wind or even snow and ice can be perilous, and every precaution taken if it is still *absolutely necessary* to climb.

If there is any possibility of weather conditions compromising the venue or stage safety the only option is to cancel the event and suspend all but vital work to make the structure safe. Contingency plans for cancellation or postponement of all or part of the show should be worked out long in advance of the event itself.

It should be noted that water and dampness could cause considerable problems for some equipment such as natural fibre ropes and fabrics. It is important to make checks for signs of rot, rust and decay following wet periods and make records of such checks.

Some years ago, I was involved with a festival held on a beach in Cornwall. There was only one access road onto a site that was otherwise surrounded by cliffs on three sides and the sea on the other, there was also a stream that issued out from the base of the cliffs at one point cutting a path through the sand dunes down to the sea. The risk assessments for this festival involved obtaining geological reports to assess the structural integrity of the cliffs and consulting tide tables. We had to put in place signs warning of the dangers of entering the sea or the stream as well as having a team of life guards on duty to protect the public that did enter the sea and stream. Fences, signs and stewards were used to warn and protect the public from the dangers associated with the cliffs and to prevent people climbing the cliffs.

A friend of mine had to do a risk assessment on giant hogweed that was found during a pre-event site visit, there are other 'nasties' such a e-coli, ringworm, lime disease, leptospirosis, tetanus, and many others that need to be thought about and the respective risks assessed.

Leptospirosis, also know as Weils disease, can also be found at indoor venues as well as on outdoor sites. I know of one case where a rigger contracted the disease whilst crawling about in the roof space of an old theatre where rats were present. It entered his system through a tiny cut on his hand, but it's usually found in bodies of water where rats are present; this includes streams, rivers, ditches, ponds, lakes and puddles. The disease is normally carried by rats as a bacterial infection in the rat's kidneys and is transmitted by the rat's urine, but cattle have also been known to be carriers. Rats are excellent carriers as unlike most animals they urinate whilst still running, thus they manage to spread urine over a greater area.

Infection is most likely to occur:

a) at the water's edge
b) after floods
c) in warmer temperatures (summer/autumn)
d) in stagnant or slow moving water

The disease can enter the body through the mucous membranes of the eyes, nose or mouth, but cuts or scratches are the main source. Unless diagnosed and treated rapidly it is often fatal with about half a dozen deaths per year in the UK, and if it doesn't kill you, you may have to give up alcohol for the rest of your life!

In the initial stages the symptoms include temperature, fever, muscular pain, loss of appetite, vomiting, exhaustion, nose bleeds, cold sores, bloodshot eyes, a rash that may go on to produce bleeding, urine colour change from light to dark, and delirium.

Tetanus is also a killer disease that is carried in dirt and can enter the body through cuts and scratches, barbed wire is a very good carrier particularly where cattle are present. Regular vaccinations are available from your GP and are recommended for anyone who does any kind of work that involves getting dirty.

Are crew are advised to stay away from streams, rivers, ponds, lakes and ditches on site as well as staying out of hedges and bushes where it may be possible to be cut, scratched or in contact with all manner of unpleasant things. To help prevent germs and diseases from propagating, the work area should be kept clean and tidy, rubbish should be disposed of correctly and not allowed to build up and create a risk by being a fire hazard.

Even the smallest cut, scratch or injury sustained on site should be thoroughly cleaned, treated with antiseptic and covered. The accident book *must* also be filled in.

Several years ago, a number of visitors to Glastonbury Festival contracted e-coli from the mud on site. To cut the chances of e-coli being present, cattle should be removed from any site at least two weeks before the event, about a week before the event the ground should be chain-harrowed to break down any cow pats that may remain, which will help to destroy any remaining e-coli pathogens.

In 2001 at a small festival in France, a tree fell to the ground killing about a dozen festival goers and seriously injuring many others who were sheltering from the rain in a tent erected under the tree. At the time the tree fell, the wind was not much greater than a strong breeze.

Regrettably this is not really a 'foreseeable' accident, but a little can be done to reduce the risk.

As part of the site risk assessment any trees on site should be examined for signs of disease, splitting, rotting, fungus and limbs bowed down with the weight of leaves. It should be established if there is any history of trees falling in the area. This information can often be found from the landowner, farmer, groundsman or gardener, and if any of these signs are showing, then a tree surgeon should be brought in to give a second opinion and carry out any work to treat and repair and make safe the tree/s in question.

24 HEALTH AND SAFETY (SAFETY SIGNS & SIGNALS) REGULATIONS 1996

These regulations replace the Safety Signs Regulations. In 2003 the Chemicals (Hazard Information and Packaging) Regulations (CHIP) were also included in these regulations that also include information on weights and dimensions.

These regulations are designed to standardise safety signage across Europe and ensure that with the free movement of labour there will be no risk of safety signs being misunderstood.

A 'safety sign' is a sign referring to a specific object, activity or situation and includes signboards, a safety colour, an illuminated sign, an acoustic sign, a verbal communication or a hand signal.

The regulations require every employer to:

- provide employees with comprehensive and relevant information on what to do in relation to safety signs, and
- ensure that employees receive suitable and sufficient instruction and training in the meaning of safety signs.

This legislation requires employers to use a safety sign wherever there is a risk that cannot be controlled by any other means.

A safety sign shall be provided and maintained by the employer where a risk assessment shows that the employer:

- cannot avoid or adequately control risks in other ways (having adopted all other preventative and protective measures), and
- there is no longer a significant risk of harm occurring.

These signs should warn and/or instruct concerning the nature of the risks and the measures to be taken to protect against them and must (with certain exemptions) comply with the descriptions in Schedule 1 of the regulations.

There are dozens of different safety signs for prohibition, mandatory safety, kitchen safety, fire and emergency, first aid, fire exits, fire equipment, access and vehicle hazards, chemical hazards, road works, etc.

All help to draw attention to Health & Safety matters, make people aware and give information. There should be more use of these signs in our work areas but they must not be used as a substitute for other means of controlling risks to employees and the public; safety signs are to warn of any remaining significant risk or to instruct employees of the measures they must take in relation to these risks.

Probably the best way to get free details on all the available signs is to obtain a catalogue of signs (and other safety equipment) from one of the large suppliers of such equipment, see your local Yellow Pages (under Safety) for details.

Trucking companies may like to note that if they transport cylinders of LPG Gas for caterers then a warning sign must be in place on the truck, but the regulations regarding the Carriage of Dangerous Goods are about to change: a licence may soon be required to carry even a single small camping gas cylinder and trucks will have to be specially equipped to carry catering gas cylinders.

I can see a time when safety signs will be much more prolific, but take note: on a site it is the contractor's responsibility to provide them to help make safe their operations!

I should also mention three common verbal signs:

- **Heads!**: This is a general warning that an item of equipment has been dropped, is about to be dropped or is being lowered from above. If you in the area don't look up but take cover, you should have a helmet on if you're in this area!
- **Truss Coming In:** Means a truss is being lowered. Again, don't look up, get out of the way and take cover, you also should have a helmet on.
- **Truss Going Out:** Means a truss is being lifted. Yet again, don't look up, get out of the way, take cover and keep your helmet on. If you don't have a helmet you should not be in the area of any work at height operations!

Other types of signs are common on sites and routes to event sites, these are normally directional, information and identification signs and form a vital part of the crowd safety and management plan, the usefulness of such signs should not be underestimated; better too many than too few.

Viewing Distance Table for Safety Signs

As a rule of thumb, viewing distances can be calculated from the measurements of the graphic symbol element contained within a fire safety sign.

Graphic Symbol Height	Viewing Distance
100mm	17.0 metres
110mm	19.0 metres
120mm	20.4 metres
130mm	22.0 metres

From any point within a building or venue persons should have sight of the nearest exit. If this is not the case a sign, or series of directional signs, should guide people to the exit.

Samples of Old and New Style Signs

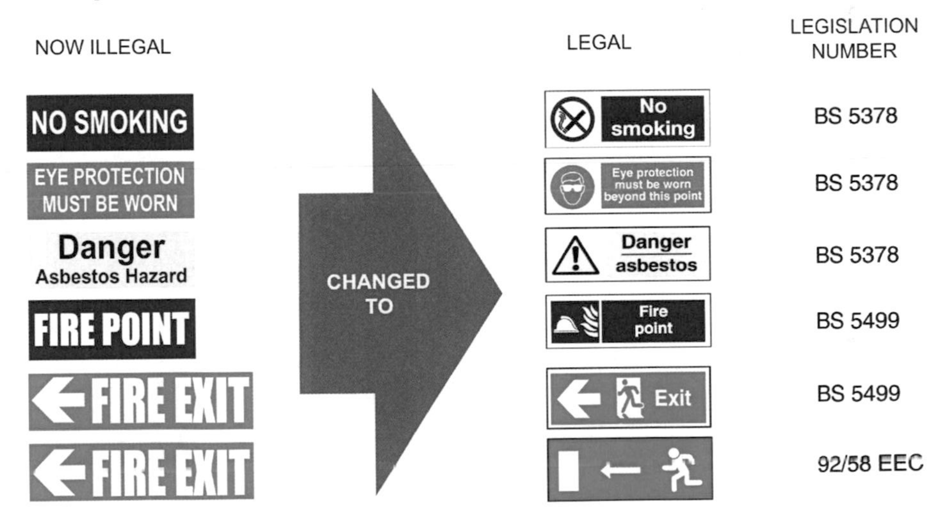

There is no such thing as an accident. They are caused by lack of control, bad planning, poor training, lack of experience, lack of risk assessment, lack of quality control and bad systems of work.

SAFETY SIGNS

FIRE POINT AND/OR EQUIPMENT

FIRST AID OR PLACE OF SAFETY

MANDATORY (MUST DO)

PROHIBITION (MUST NOT)

HAZARD

CORROSIVE

HIGHLY FLAMMABLE

DANGEROUS FOR THE ENVIRONMENT

HARMFUL/IRRITANT

EXPLOSIVE

OXIDISING

(VERY) TOXIC

25 PERSONAL PROTECTIVE EQUIPMENT AT WORK REGULATIONS 1992 (PPE) & THE CONSTRUCTION (HEAD PROTECTION) REGULATIONS 1989

The definition of personal protective equipment is very broad and includes all equipment (including clothing affording protection against the weather) which is intended to be worn or held by a person at work and which protects him/her against one or more risks to his health or safety, and any addition or accessory designed to meet that objective.

Employers are required:

- To ensure that suitable personal protection equipment (PPE) is provided to employees who may be exposed to a risk to their health and safety while at work (unless some other form of protection is already in use and is equally or more effective) [Regulation 4].
- To take into account the ergonomic requirements of the person who is to wear the PPE. It must fit the wearer correctly, be effective and must comply with any Community standards [Regulation 4(3)].
- Where more than one piece of PPE is worn at one time, that all such equipment is compatible and effective [Regulation 5].
- To carry out an assessment of any PPE before it is provided [Regulation 6].
- To keep PPE in good working order [Regulation 7].
- To store PPE when not in use [Regulation 8].
- To provide comprehensive instruction, training and information on the risks which the PPE will limit, instructions for use and how to keep the PPE in good working order [Regulation 9].
- To ensure that the PPE is used properly [Regulation 10(1)].

The self-employed also have an obligation to ensure that they also provided themselves with suitable PPE and the regulations listed above, where relevant, apply to the self-employed.

In relation to **Regulation 10**(1), employees are under a corresponding obligation to use any PPE provided to them in accordance with the training and instructions which have been given to them [Regulation 10(2)]. In addition, employees who have been provided with PPE are also under a duty to report to their employer any loss of or obvious defect in any equipment provided [Regulation 11].

As we have already concluded, Personal Protective Equipment and clothing should only be considered as a 'last resort' means of control. All other options such as engineering controls or Safe Systems of Work should be considered and used as a first choice. The reason for this is simple: other methods of control are *preventative* controls, PPE is *protective*.

Even where these controls have been applied some hazards may still remain, they include injuries to:

- The lungs, e.g. from breathing in contaminated air
- The head and feet from falling materials
- The eyes from flying particles, splashes of corrosive liquids, lasers and welding torches
- The skin from contact with corrosive materials
- The body from extremes of heat or cold.

PPE *is* required in these cases to reduce the risk.

Earplugs and earmuffs are covered under the Noise at Work Regulations but the following information will still apply.

The law says: "An employer *must* provide any required PPE free of charge to his/her employees", and that: "Self-employed persons *must* provide their own PPE". An employer may make a charge to his employees for any required PPE if the employees take the PPE home and uses it outside of work, for instance if an employee is provided with safety footwear and the employee then uses it for 'gardening' at home.

It also goes on to say that employers and self-employed persons are required to maintain (including replace or clean as appropriate) any PPE provided to his/her self or employees in an efficient state, in efficient working order and in good repair. Helmets and other items of PPE must be replaced by an employer (or self-employed person) at regular intervals, as recommended by the manufacturer.

Storage accommodation for employees' PPE when not in use *must* be provided by an employer.

Regulation 10(1) of the PPE Regulations states that every employer shall take all reasonable steps to ensure that any PPE provided is used properly after instruction and training in its correct use has been given (Regulation 10(2). The self-employed are required to obtain suitable and adequate training in the correct and proper use of any required PPE.

Other than ear protectors, the main items of PPE we are going to come across are safety footwear, helmets, fall arrest equipment, gloves, weather protective clothing and sometimes goggles or eye shields and masks. Caterers will need to have items such as oven gloves, rubber household gloves, aprons, etc.

For the majority of manual handling operations we are involved in, I would regard gloves as a matter of personal choice. However, if there is a risk of cuts, burns, abrasions, contact with certain hazardous chemicals, or contact with litter and waste (including waste from portable toilet units), then they **must** be used. A high standard of personal hygiene must also be maintained by those removing litter and waste to help prevent any kind of infection.

There is absolutely no reason why protective footwear is not 100% standard for anybody involved in heavy manual handling in our industry.

This does not necessarily mean heavy boots, there are steel toe capped trainers, shoes and Wellington boots available. Protective footwear is also a requirement within any Hard Hat Areas.

There is an even greater reluctance to use safety helmets but they must be used when lifting or moving objects over and above head height (this may include when loading/unloading a truck), when men are working overhead, when there is a risk of falling objects and when risk assessment dictates. At these times the area must be cleared of all unnecessary persons and a Hard Hat Area established.

For serious impacts, helmets are a once only device; they should be discarded and destroyed after serious impact or if dropped from a height since they will not protect against a second impact and your life may depend on this. Helmets should be stored at room temperature when not in use and out of sunlight since the polymers used in manufacture are affected by sunlight and ultra violet light so don't leave them on the parcel shelf of your car. Helmets should not be modified or have holes drilled into them as this may alter the impact protection provided.

Keep helmets clean and free of dirt and wash with a mild detergent in water. Chemicals such as glues, paints, oils and solvents should also be kept away from helmets.

It is impossible to give an expected usable life for a helmet; it depends on use and it is possible to irreversibly damage a helmet on its first outing. A helmet made from composite materials may have a life of 10 years if used correctly, a helmet made from thermoplastic polymers has a maximum life of four years but this may be as little as one year in certain circumstances. There is now a large selection of safety helmets to choose from including lightweight ventilated helmets and low profile helmets without a peak that do not obstruct upward vision.

Riggers and those involved in work at height operations should be aware that the normal safety helmet is only designed to protect the user only from falling objects; it offers little or no protection to the user should he or she take a fall. At these times a helmet of the type used by rock climbers must be used. It should carry a CE Mark and comply with EN 136.005.05 and the UIAA standards. Furthermore, the law requires that a worker uses a harness and a fall arrest system *at all times* when working at a height where a fall could cause injury – this could be a little as six inches! (See 'Work at Height').

Generator fitters and engineers will need gloves and overalls to protect them from diesel and oil contamination. They should be made aware that oil or diesel soaked rags should not be kept in overall pockets, there is more than one case on record of mechanics developing dermatitis or even cancer of the testicles from this bad practice. Barrier cream must be used and is yet another item of PPE.

Extremes of weather conditions should not be overlooked and suitable PPE obtained, this may consist of waterproof jacket (with hood), over trousers, boots, gloves, hat and if required, thermal underwear.

Staging companies and others working outdoors should consider the danger of skin cancer when working in the sun and staff are advised to cover up exposed areas in particular the head, neck, back, arms and legs. Sun block may need to be added to the list of PPE.

A supplier of Personal Protective Equipment can advise on the correct and most suitable PPE for the operation in question.

PPE must be correctly fitting and suitable for the purpose intended. It must also meet the required British or new European Standards. Equipment meeting these standards will display the BS Kite Mark or the European CE Mark, it is essential that equipment of this standard is provided; an employee who crushes his foot while using boots that do not meet the required standards has a very strong case against his employer!

The British Standard 'Kite Mark'

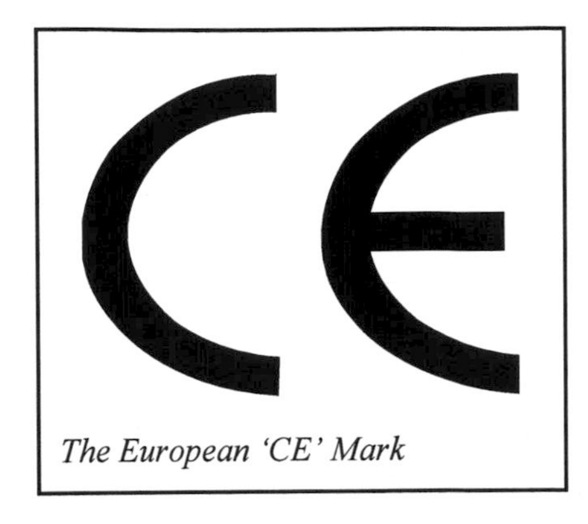

The European 'CE' Mark

CE marking is in fact a group of marks that must appear on the product or its labelling to enable its free legal movement across borders, its sale and/or use within the EEA (European Economic Area). The letters CE have to be in typeface as shown above and this is usually followed by other information. On PPE these marks show the equipment meets all the requirements of the PPE directive.

Since July 1995 it has been illegal for manufacturers to sell new equipment without the CE mark in the EEA. Eventually, all PPE on the market will carry CE marking but if the product is very small the marking will be on the packaging.

There are three categories of PPE but only two of these are named and defined by the directive (simple and complex).

I Simple – e.g. washing up gloves
II Intermediate – e.g. all items that are neither simple nor complex
III Complex – e.g. harnesses and fall-arrest equipment

Category III equipment must:

1. Undergo independent type testing, usually to European Standards (Ens).
2. Have appropriate technical and user instructions.
3. Be produced under an independent verified quality system, e.g. ISO 9000 (BS 7505) or subject to periodic independent testing.

Category I equipment does not require conditions 1 and 3.

Category II equipment does not require condition 3.

The requirement for CE marking is currently:

Category	Requirement	Example
I	CE Mark only	CE
II	CE Mark plus year of affixation	CE95
III	CE Mark plus year of affixation plus number of notification body controlling the manufacturer's quality system.	CE95 0456

When selecting equipment, ensure it meets the standards relevant to its intended use. These could be European (Ens), International (ISO) or national, such as British (BS), German (DIN), or French (AFNOR).

The best thing to do is look for compliance with European standards. If these standards (Ens or PrEns) do not exist, then look to some of the other standards listed here.

If you are not sure if a particular standard is relevant to your intended use contact the manufacturer, they will usually be able to offer advice.

Manufacturers achieve CE marking by meeting the harmonised European Standards (European Norms or Ens), where they exist. The European Committee prepares these standards for standardisation (CEN). Member states of the EEA can participate in the production of standards via their national standards' body, e.g. the British Standards Institute. If a CEN standard does not exist there are other ways that a CE marking can be obtained.

As harmonised standards are introduced they are adapted by each member state and become dual standards. For example, the standard for a full body harness is BS EN 361; in Germany this is shown as DIN EN 361 etc. Throughout Europe, national standards which conflict with or overlap these new CEN standards, will be withdrawn.

There are heavy penalties for the abuse or misuse of anything (including signs and PPE) provided for safety. A hard hat is not a football, so don't kick it around.

Whilst not strictly PPE, the right clothing can make the job far more comfortable and make crew look far more professional. Black clothing is most suited as it helps crew 'blend in' with drapes and backdrops so they are not so obvious to the audience, not only that but an added bonus is that black does not show the dirt so much so you don't have to find a launderette so often when on tour.

Jeans, T-shirts, sweat-shirts and tracksuit trousers are quite suitable so long as they are a good fit, do not restrict movement and do not have flapping or dangling parts that may catch in equipment or machinery. By the same token, jewellery, wristbands and laminate passes should not be worn and long hair must be tied back if the operation so requires for safety reasons.

Whilst not a legal requirement, it is strongly advised not to share items of PPE, particularly items such as earplugs, gloves and boots; this will help stop the spread of infections, etc.

Other items such as harnesses should not be shared as the continual adjustment of straps does not do a lot of good to the harness. PPE should be issued new to an employee who then becomes responsible for its safekeeping.

Where visitors are expected on site it may be necessary to have a stock of helmets, etc. that visitors may use. Visitors must be accompanied at all times by a competent person who will be aware of the use given to visitor PPE and who can withdraw damaged or worn items. After use, items such as visitor helmets should be cleaned and disinfected with an antiseptic wipe before being correctly stored.

The Construction (Head Protection) Regulations 1989 states:

"Everyone has a role to play to in making sure that head protection is worn when necessary." [Regulation 4]

Employers and those who have control over people at work on operations or works, have a duty to ensure suitable head protection is being worn unless there is no foreseeable risk of injury to the head other than by falling. Employees and the self-employed have duties to wear head protection. *(A normal type safety helmet will not protect the wearer's head in the consequence of a fall; this type of helmet is only designed to protect the head from falling or swinging objects and from striking the head against something - particularly if there is insufficient headroom.* [Regulation 4]

Employees must wear head protection when directed to do so by their employer or to comply with written rules made by the person for the time being having control of the site/venue, this could be the promoter, production manager, stage manager or site manager.

Self-employed persons working on their own, i.e. not under the control of another person, must wear head protection unless there is no foreseeable risk of injury to the head other than by falling (in which case a 'climber's' helmet is required).

Every employee who has been provided with a helmet by his employer must take care of it and report any loss or defect to their employer as soon as possible. Any concerns about the serviceability of a helmet must also be reported.

Regulation 5 states that the person in control of a site (promoter, production manager or stage manager or site manager) may (so far as is necessary to comply with Regulation 4) make rules regarding the wearing of helmets on site. These rules must be made in writing (after consultation with employees) and all staff (including union and safety reps.) informed.

Rules should be clear and unambiguous to ensure there is no misunderstanding or doubt about the need to wear head protection.

Those with duties to ensure suitable head protection is worn (and this may also include Crew Chief's) will need to:

1. Identify when and where head protection should be worn
2. Inform personnel
3. Provide adequate supervision
4. Check head protection is worn when necessary

Information can be given by erecting safety signs informing that head protection should be worn. This is also a good way of reinforcing the 'wear your head protection' message. Safety signs must conform to the Health & Safety (Safety Signs and Signals) Regulations 1996.

Supervision by those responsible (again, this includes crew bosses) for ensuring head protection is worn should be a habitual activity – including looking out for helmet use at all times, starting early in the day and taking in later arrivals to the site or venue.

Employers must satisfy themselves that all necessary action has been taken to comply with their duty under the regulations to ensure head protection is worn by their employees.

All those with a measure of control over the work must also satisfy themselves that action has been taken, to comply with their duty under the Regulations.

Experience has shown that those who have been trained to use PPE correctly gain far better protection from the PPE than those who have not received the same training. A good example is in the use of earplugs: just stuffing the plugs into the ears offers little or no protection, yet those who have been trained to fit the plugs correctly gain a very high level of protection. This highlights yet again the need for proper training within the industry.

The HSE has issued a warning to people who work with harmful dusts that they should not use 'nuisance' dust masks to protect themselves from exposure.

The HSE is encouraging the voluntary withdrawal from sale of nuisance dust masks, also known as comfort masks or hygiene masks, and is urging the use of approved CE-marked disposable respirators instead.

The warning is in support of the Health & Safety Commission's (HSC) campaign to reduce respiratory diseases such as occupational asthma, which is the most frequently diagnosed occupational related respiratory disease in Great Britain. The HSE estimates that between 1,500 and 3,000 people develop it every year.

Nuisance dust masks should only be used when dusts are not hazardous to health, but they are often used as cheap alternatives to the correct protective equipment.

Despite resembling the kind of disposable respirators that should be used, nuisance dust masks are not protective devices. They perform badly and do not have to meet basic Health & Safety requirements. However, they are readily available to industry, DIY enthusiasts and hobbyists, which means that they may be found in a variety of occupational and domestic environments and consequently could be used for purposes for which they are not intended.

They should not be used for protection against fine dusts, welding fumes, asbestos, fine sand, paint spray, gases, vapours or aerosols. In addition, they should not be used for substances with a maximum exposure limit because the law requires that exposures to these be reduced as much as possible.

This means they are unsuitable for protection against grain dust, flour dust, ferrous foundry dust, hard or softwood dust, wool process dust or fume from rosin-based solder flux.

Similarly, nuisance dust masks should not be used to give protection from substances that cause occupational asthma but do not have occupational exposure limits. Examples include some dyes, antibiotic dusts, proteolytic enzymes and mould spores.

The HSE will enforce against employers who issue nuisance dust masks inappropriately when proper respirators should be used, although the HSE cannot prohibit their sale. A statutory ban on these masks would require a change in European law. However, the HSE is committed to ensuring that workers' health is better protected.

In particular, the HSE has set a target of 30% reduction in new cases of occupational asthma by 2010. If nuisance dust masks have been used because of a lack of proper risk assessment, replacement by appropriate disposable respirators should show health benefits.

Nuisance dust masks may consist of a thin metal plate that holds a piece of gauze over the nose and mouth or a lightweight filter that looks similar to a disposable dust respirator. Nuisance dust masks are not classified as personal protective equipment.

They are not CE-marked to show that they comply with European Directive (89/686/EEC) or against a recognised standard for respiratory protection.

All disposable respirators must be individually CE-marked. They are also marked to show the European standard: EN 149, and class: FFP1 (low efficiency); FFP2 (medium efficiency); or FFP3 (high efficiency). EN 149 indicates that the respirator conforms to British Standard EN 149:2001 Filtering half masks to protect against particles.

Employers are reminded that any form of mask or respiratory protective device is a last resort, and that, wherever possible, dusts or chemicals should be controlled by other means such as enclosure and local exhaust ventilation in line with the Control of Substances Hazardous to Health (COSHH) Regulations 2002.

You can't just delegate responsibility – you have to delegate authority as well. Having the responsibility is no good if you don't have the authority and having authority usually means having control of some or all of a budget so that safety controls can be implemented.

26 MANUAL HANDLING OPERATIONS REGULATIONS 1992

These regulations (as Amended in 2002) apply to occupations not previously the subject Health & Safety legislation. Past legislation was limited to work in defined locations such as factories, shops and offices, whereas these new regulations cover occupations as diverse as nurses, delivery drivers, university technicians, rock 'n' roll concert and event production crews.

As human beings, we are obsessed with moving things about. Approximately 34% of all accidents reported each year to the enforcing authorities are associated with manual handling – the transporting or supporting of loads by hand or by body force. 65% of all manual-handling injuries are either sprains or strains, and 45% of all manual-handling injuries are located in the back. There are approximately 50,000 manual-handling accidents per year, that's 200 per day! It's not clever or macho to be able to lift three monitor wedges (or similar) at a time!

Let's first summarise the most common types of manual-handling injury:

Back Injuries

Caused by twisting, lifting or pushing loads, which are borne on the spine.

Between the vertebrae (the bone structural joints of your spine) are fluid-filled cushions of gristle called intervertebral discs. Excessive torsoinal or crushing movement on the spine can lead to displacement of the fluid, causing painful backache, lumbago or sciatica. This condition is also known as a 'slipped disk'.

Muscular Sprains and Strains

When a muscle is stretched beyond its normal limit it is strained. A sprain is caused by sudden or excessive force, thereby weakening the joint and related muscles.

Hernia

Usually a consequence of incorrect lifting or lifting a heavy load. The muscalature of the lower abdomen is strained to the point where a rupture of the body cavity wall takes place, allowing a protrusion of part of the intestine.

Cuts, Abrasions and Bruising

The risk of such injuries can often be reduced by wearing the appropriate protective clothing.

Crushing

Fractures, broken or cracked bones are often the result of jamming fingers or dropping objects on feet.

Repetitive Strain Injury

Mention must be made of Repetitive Strain Injury (RSI), caused by work which demands a constant 'unnatural' movement or posture.

It can lead to pain or restricted movement in the afflicted limb(s) or torso, with muscle fatigue, joint dislocation or fracture. Even office workers can be affected, for example, by a badly designed chair or poorly positioned computer keyboard. Postural problems of this sort are usually caused by bad ergonomic design. Therefore, management must be aware of good practice.

The Manual Handling regulations apply to the handling of loads by human effort, as opposed to mechanical handling by crane, lift truck, etc. The human effort may be applied directly to the load or indirectly by hauling on a rope or pulling a lever. Introducing mechanical assistance, for example a sack truck or a powered hoist, may reduce but not eliminate manual handling since effort is still required to move, steady or position the load.

Manual handling includes both transporting a load and supporting a load in a static posture. The load may be moved or supported by the hands or any other part of the body, such as the shoulder. Manual handling also includes the throwing of a load, whether into a receptacle or from one person to another.

The application of human effort for the purpose, other than transporting or supporting a load, does not constitute a manual-handling operation. For example, lifting a control lever on a machine or the action of pulling on a rope whilst lashing down a load on the back of a vehicle is not manual handling.

Employers are required:

- As far as is reasonably practicable to avoid MHO where there is a risk of injury. Regulation 4(1)(a).
- Where it is not reasonably practicable to avoid MHO, to make an

assessment based on the factors set out in the Schedule to the Regulations.

There are four factors: the task, the load, the working environment, and individual capability.

- To take steps to reduce the risk of injury to the lowest level reasonably practicable, and where it is reasonably practicable, to give precise information on the weight of the load and the position of the centre of gravity if it is not located centrally in the load. Regulation 4 (l)(b)(ii).
- To review the assessment if there is any reason to suspect that it is no longer valid, for example, if any injury were sustained while carrying out that manual operation. Regulation 4(2).
- Employees are under a duty to make full use of any system provided by an employer after an assessment.

The Manual Handling Operations Regulations affect most of us working at events. It again requires employers to complete written risk assessments.

If the operation can be avoided or mechanised by the use of a forklift truck or other mechanical lifting aids, such as a set of sack trucks, then this should be the first course of action. Unfortunately, manual handling is something that generally can't be avoided in our industry.

Regulation 4 of the Manual Handling Regulations requires the employer to reduce the risk of injury to employees involved in manual handling operations to the lowest level reasonably practicable. Employers are therefore required to train staff in safe manual handling techniques and provide Personal Protective Equipment (PPE) such as safety footwear, gloves and, where loads have to be lifted over and above head height, safety helmets. (The self-employed must supply their own).

As a simple risk assessment will prove, if a load when accidentally dropped is capable of doing damage to the feet of the operator, then safety footwear must be worn. You have no excuse!

Regulation 14 of the Management of Health and Safety at Work Regulations 1999 requires employees to make use of equipment provided for them in accordance with the training and the instructions their employer has given them or provided for them.

Because of the vast and varied nature of the loads we are required to move and variations in each and every venue/warehouse, it is probably almost impossible to write anything other than a generic risk assessment for our manual-handling operations.

Remember: assessment is not an end in itself, merely a structured way of analysing risk and pointing the way to practical solutions. Any duty imposed by these regulations (and in fact, most of the other regulations) on an employer in respect of his employees shall also be imposed on a self-employed person in respect of himself.

Within the regulations are factors to which the employer must have regard and questions he must consider when making an assessment of manual-handling operations. They are as follows:

The Tasks. Do They involve:

- Holding or manipulating loads at distance from the trunk?
- Unsatisfactory bodily movement?
- Twisting the trunk?
- Stooping?
- Reaching upwards?
- Excessive movement of loads, especially excessive lifting or lowering distances?
- Excessive carrying distances?
- Excessive pushing or pulling of loads?
- Risk of sudden movement of loads?
- Frequent or prolonged physical effort?
- Insufficient rest or recovery periods?
- A rate of work imposed by a process?

The Loads. Are They:

- Heavy?
- Bulky or unwieldy?
- Difficult to grasp?
- Unstable, or with contents likely to shift?
- Sharp, hot or otherwise potentially damaging?

The Working Environment. Are There:

- Space constraints preventing good posture?
- Uneven, slippery or unstable floors?
- Variations in level of floors or work surfaces?
- Extremes of temperature or humidity?
- Ventilation problems or gusts of wind?
- Poor lighting conditions?

Individual Capability. Does the Task Require:

- Unusual strength, height, etc.?
- Special information or training for its safe performance?
- Does the job put at risk those who are pregnant or who might reasonably be considered to have a health problem?

Service companies can assist further by marking flight cases with their approximate weights, centre of gravity and other warnings such as 'Keep This Way Up', 'Do Not Tip', 'Unstable Load' or 'Contents Liable To Move', as well as making sure that cases are in good condition with sufficient handles, no sharp edges, and smooth running wheels or wheel dollies. In fact the regulations state:

Each employer shall:

Where it is not reasonably practicable to avoid the need for his employees to undertake any manual-handling operations at work which involve a risk of being injured, take appropriate steps to provide any of those employees who are undertaking any such manual-handling operations with general indications and, where it is reasonably practicable to do so, precise information on the weight of each load, and the heaviest side of any load whose centre of gravity is not positioned centrally.

I can think of very few companies within our industry that carry out any manual-handling training and therefore has the evidence and records to show that the staff has been trained.

In this situation, any member of staff suffering even a minor back injury could cost his employer hundreds of thousands of pounds if he chose to take legal advice – remember the no win-no fee injury claims companies?

Some basic safety rules apply to manual handling:

- Lifting and moving of objects should always be done by mechanical devices rather than manual handling wherever reasonably practicable. The equipment used should be appropriate for the task at hand.
- The load to be lifted or moved must be inspected to see if it is stable, if the centre of gravity is not positioned centrally or is liable to change, if it is hot, unstable, if there are sharp edges, silvers and wet or greasy patches.
- When lifting or moving a load which is hot or has sharp or splintered edges, gloves must be worn. Gloves must be free from oil, grease or other agents which might impair grip.

- The route over which the load is to be lifted or moved should be inspected to ensure that it is free of obstructions or spillages which can cause tripping or spillage.
- No one should attempt to lift or move a load that is too heavy to manage comfortably.
- Where team lifting or moving is necessary one person should act as coordinator, giving commands to lift, lower etc. Do you lift 'on three' or 'after three'? It is often useful to have this person stand at a short distance from the operation so that he may obtain a full view of the operation.
- Members of a team carrying out a 'team lifting' operation should be of a fairly equal size and stature. It's no good having a mix of midgets and giants. In most situations a 'humping' crew should be an even number and not an odd number of persons.

When lifting an object straight off the ground, assume a squatting position, keeping the back straight. The load should be lifted by straightening the knees, not the back. These steps should be reversed for lowering an object to the ground.

Where you have to handle bulky objects in high winds you may have problems trying to maintain your balance. You may need to provide windbreakers. Some operations such as fitting or removing the sheeting or scrims on an outdoor stage will only be safe when there is little wind, and when working at height the operator must be secured by a shock-absorbing lanyard and full body harness (See Work at Height). Operations such as these need extra care, so make sure you are securely fastened on so you can't fall.

It's no good putting it on paper if you don't put it into practice!

Don't just 'pay lip service' to health and safety.

27 THE NOISE AT WORK REGULATIONS 1989

The Noise at Work Regulations 1989 came into force in January 1990 and placed duties on employers, employees, and people who supply and use noisy machinery and equipment. These regulations were introduced with the aim of protecting the hearing of workers in noisy working environments and have nothing to do with the noise levels set by the conditions of an Entertainment Licence. The latter are put in place to reduce possible disturbance to those neighbouring the licensed premises. In effect, we often have two sets of levels to deal with: inside the venue and environmental noise problems caused by noise 'leaking' out of the venue.

For all staff (even volunteer staff are classed as employees under Health & Safety law), contractors and sub-contractors, all festival sites, concert halls and all other venues are by law places of work and are therefore governed by the regulations. We may be reluctant to come to terms with this fact, but as with all the other regulations, we are now stuck with them.

We risk high fines and possible imprisonment if we do not comply. I must admit that, apart from at a few major festivals, I have never seen this regulation enforced, even in local authority owned venues (but that is not to say that it is not enforced). To the best of my knowledge, at the festivals where it has been enforced the enforcement has been limited to establishing the 'pit area' as an ear protection zone and ensuring that ear protection was available and used by those who were required to enter the pit area. This is all well and good and I can hear you rubbing your hands together with glee because nobody is enforcing the law, but what happens in twenty years' time when a technician tries to sue the band he was working for back in 1999 because now, in 2019, he has developed a serious hearing problem and his employer (at the time he thinks the damage was caused), did not warn him officially of the hazard, supply him with ear protectors, and introduce other control systems to protect him?

The Hazard

This is obvious: *Noise*. But the source of the hazard may not be so obvious, thus a little thought should be given to non-obvious noise sources.

PA systems are an obvious noise source, but this is where things do get a little complex. Here we are trying to create what is technically a hazard for people to expose themselves to for the purpose of enjoyment. The law says we should try to remove the hazard or at least reduce the risk to protect people!

Noise (as we all know) is measured in decibels (dBA), but sudden impact noises produced by hammer blows, percussive drills, pyrotechnics, post bashers, etc. are measured in Pascals, a measure of air pressure. These sources are also present at many of the sites and venues we work at.

The Risk

Loud noise can cause irreversible hearing damage. It accelerates the normal hearing loss which occurs as we grow older and it can cause other problems such as tinnitus and interference with communication. It can lead to the increase of other accidents and stress.

An Indication of Noise Levels

As a 'rule of thumb guide', if you cannot clearly hear what someone is saying when you are around two metres away, the level is likely to be around 85dB(A) or higher, and if you cannot hear someone clearly when you are about one metre away, the level is likely to be around 90dB(A) or higher.

Where PA systems are installed it is certain that noise levels at the stage, mixer positions, pit, and areas around the speaker systems will be above 90dB(A), even behind the speakers. These levels may well extend much further.

The noise from the average forklift truck is about 101dB(A) and about 117dB(A) from an un-silenced generator. Noise from fireworks is normally measured in Pascals.

But What is Loud?

Noise Comparisons dB (A)

0	Hearing threshold
10	Rustling leaf
30	Human whisper
60	Normal conversation
80	Busy traffic
90	Heavy goods vehicle
100	Engineering workshop

110	Angle grinder or average rock concert
120	Prop-driven aircraft
130	Riveting hammer
140	Jet engine aircraft

Important: the decibel scale is logarithmic. An increase of 3dB represents a doubling of the actual noise level. I heard of an Environmental Health Officer who thought that to reduce the volume by 50% from 100dB(A) you had to bring it down to 50dB(A): not a very bright example of an EHO!

The Law

The Noise at Work Regulations 1989 are intended to reduce hearing damage caused by loud noise. There are three noise levels at which the regulations state that action needs to be taken by employers to protect their employees from the harmful effects of noise. The first action level means a DPNE of 85dB(A). The second action level means a DPNE of 90dB(A). The peak action level means a level of peak sound pressure of 200 Pascals.

What We Must Do

All employers have a general duty to reduce the risk of hearing damage to their employees to the lowest level reasonably practicable. This may be done in numerous ways. For instance:

1. Can machinery and equipment be modified to reduce the level of noise generated?
2. Is alternative, quieter machinery/equipment available?
3. Do all the affected workers need to be in the noisy area or could some jobs be done in other, quieter areas?
4. If it is not possible to reduce noise levels by any other means, then hearing protection should be considered, but only as a last resort.

We must first identify the areas for concern. Many of these will be quite obvious, such as on and around stages with PA Systems or on items of plant and machinery like tractors and forklift trucks. Employers have a duty to reduce the noise level to 85dB or below (the first action level).

We must also identify areas (a basic dB metre may be useful to help identify the less obvious sound sources) within the first and second levels of action and mark off these zones. Mandatory signs must be erected to identify these zones to others. The signs must be in accordance with the Health & Safety (Safety

Signs & Signals) Regulations 1996, (white on a blue circle).

For a tour, concert or festival situation, it will again fall back to the promoter, venue or production manager, depending on circumstances. Festivals may well have Health & Safety Advisers or Consultants whose role it is to carry out these assessments. Whoever does the assessment, he or she must be competent and a copy of the assessment must be kept on file until a new assessment is made. A 'competent person' to carry out a noise assessment is someone with the Institute of Acoustics Certificate of Competence in workplace noise monitoring, or a higher qualification in acoustics. This will normally be a specialist consultant.

The assessment should identify who is at risk.

Provide the assessor with sufficient information to facilitate compliance with the regulations.

Having had the noise assessed, if it is found to be at the *First Level of Action*, employers must:

- Reduce the noise level to below 85db if it is reasonably practicable to do so, if it is not, the employer must:
- Tell all workers (and contractors) about the risks and the precautions.
- Make hearing protection freely available to those who want it where levels exceed 85dB(A)
- Suggest to workers (and contractors) that they seek medical advice if they think their hearing is being affected.

If after the assessment it is found that noise levels exceed 90dB(A), the *Second Level of Action*, employers must:

- Do all they can to reduce exposure (other than providing hearing protection), such as by engineering control (e.g. by fitting silencers to machinery, changing the environmental layout of the work area, dampening, isolation, appropriate maintenance), or rotating staff positions, thereby reducing the time they need to spend in noisy positions.
- In many cases, such as on and around a stage, it will not be possible to reduce the noise. Therefore, the only control will be ear protection. Drivers and operators of noisy plant and machines must wear hearing protection. Ear protection zones around stages, speakers and generators must be established, marked out and entry controlled to these zones. (See 'Enforcement', below).
- Employees (and self-employed persons) have a duty to use the ear

protection provided and report any defects or problems to employers. A peak action level of 200 Pascals (approximately 140dB(A)) is quite likely to be produced by percussive equipment and tools. Staff using or working near such equipment must use the special ear protection required for this type of work. This protection must be used even if the level is only exceeded for a few seconds.

Record Keeping

Employers must keep an adequate record of the noise assessment and its findings until the assessment is reviewed.

Reviewing the Assessment

The noise assessment should be reviewed and updated whenever there is reason to suspect that the current assessment may no longer be valid or if there is a significant change in the nature of the work to which the assessment relates.

Employees' Duties

1. To wear ear protectors in any area where the second action level is reached and any time you are working in a designated ear protection zone.
2. To ensure that all protective equipment provided is properly use and looked after.
3. To report any defective equipment to your employer.

Equipment Suppliers' Duties

A person or company supplying any equipment or machinery which is likely to expose people using it to the first action level or above must supply with it information on the likely noise levels.

Hearing Protection

There is no real substitute for removing or reducing the noise at source to protect hearing (preventative action), but as this is not possible in most cases, hearing protection (protective action) is second best.

Ear protection must be worn at all times that you are at risk. It's a bit like going out in the rain but only wearing a raincoat for some of the time you are out in the rain – you are still going to get wet. There are two main types of hearing protection:

Earplugs: either disposable or reusable.

Both types must be fitted correctly with clean hands, as dirt may cause irritation. It is a fact that people who have be trained to use and fit ear protection properly get far better protection then those who have not received any training.

There is more to it than just stuffing some plugs in your ears: just doing this will do very little good. To work correctly, they must be fitted properly, so training is essential.

The reusable type need regular and careful washing and the initial supply and fitting must be by a trained person. Disposable ear plugs must not be reused.

Ear defenders or muffs: if damaged or not fitted/worn correctly they should not be relied upon to protect hearing to the level required.

Ear protectors of any kind should not be shared amongst staff as this may rapidly spread ear infections. Anyone with an ear infection should use muffs not ear plugs.

Perhaps it comes as no surprise that very few people know how to use hearing protection correctly. Unless it is used correctly, the protection provided will offer little or no protection.

Simply sticking a pair of disposable earplugs into your ears is of little use; correct training in the use of ear protection is required. Employers (and the self-employed) have a duty to ensure the correct training is provided or obtained. A training session on the use of ear protection will only take a matter of minutes, but is essential in order to obtain the full level of protection that can be afforded to the user by the use of the correct ear protection.

Exemptions

There is HSE guidance for the various times a worker can be exposed to various levels of high volume. *There are no exemptions.* Ear protection must be worn by all persons entering the ear protection zones, even if they do not stay long enough to receive an exposure of 90dB(A) $L_{EP.d.}$

Musicians and sound engineers will find it extremely difficult to work with standard earplugs in place, but there are special plugs that reduce the sound level without altering tones and frequencies. These are often very expensive or need to be custom-made. Even though the law does not officially allow it, I would personally make an exception to sound engineers (but only when they are actually mixing) and an exception to the musicians, but please remember that this is my personal view and I am not suggesting you should follow my example as you may still get prosecuted. At the time of writing the HSE is consulting with the entertainment industry over this very problem. The regulations are to be revised and updated in 2006 to meet with new European Directives.

All other persons must wear ear protection on stage, in the pit area, the FOH mixer positions and other designated Ear Protection Zones, or when operating noisy plant and machinery.

Enforcement

The local authority will be responsible for enforcement at most events. Like all breaches of Health & Safety law, prosecution can lead to high fines or even prison sentences.

Stewards should be fully briefed so that they can help 'police' Ear Protection Zones. These zones must be clearly marked with tape or by other means and mandatory warning signs erected. Protectors can be supplied in dispensers and these should be placed at the entrances to all Ear Protection Zones. Staff should be warned of the problem and instructed on the correct use of ear protection. If in any doubt, wear ear protection.

The Public

There is no legislation setting noise limits for audience exposure to loud noise.

The conditions of most entertainment licences require that speaker systems are at a minimum of three metres from the public and that levels do not reach 140dB(A). It was considered highly unlikely that volumes would get anywhere near this level from PA systems but 143dB(A) has been recorded from audience applause at a festival!

Promoters and organisers should issue a warning to the public on all tickets and with signs at the venue/site stating that 'Exposure to high sound pressure levels can damage hearing'. There is nothing to stop a member of the public making a civil claim against a band, promoter, PA company or even a sound engineer if they can prove they suffered hearing damage at a particular gig as a result of high sound pressure levels. In fact, successful claims have already been made against PA companies and sound engineers.

New Regulations

A revised edition of the European Physical Agents (Noise) Directive came into force in February 2003. The new directive is far more detailed than the previous one and is far more prescriptive in the areas that are covered by the earlier directive.

The directive must be introduced into UK law by early 2006, but the music and entertainment sectors – including bars and nightclubs – will have until early 2008 to comply with the new exposure and action limit values for their employees. For a legal document the directive is relatively concise and its core is in articles 3 to 10; articles 1 and 2 are definitions.

Under **Article 3** of the new directive, the First Action Level (Value) must be reduced to 80dB(A), a reduction of 5dB, and reduce the Second Action Level (Value) to 85dB(A), again, a reduction of 5dB. This will mean that all members of an orchestra, playing acoustically with no amplification whatsoever, will be required to wear ear protection during their performance if they are to comply with the Law. Under the new regulations, Action Levels will become known as Action Values.

The Third Action Level (Value) must be reduced from 200 Pascals to 140 Pascals. At this level, employers have to put in place controls to reduce noise levels and ensure that all employees are wearing ear protection. In addition, a new limit of 87dB(A) is to be introduced that requires employers to ensure that weekly noise exposure – taking into account the level of ear protection offered by PPE such as ear defenders – does not exceed this limit during an 8 hour working day.

Article 4 details and reinforces the need for risk assessment, such as where there is a consideration for purchasing quieter machinery. The requirement to record the risk assessment measures time taken to control the risk, and also the need to update the assessment on a regular basis. This may include the time to consider the supply of new quieter machinery. Do you have a purchasing policy, which includes an assessment of noise risk and includes positive discrimination in favour of low noise emission machinery?

Article 5 details the normal hierarchy of controls that are normal to control any risk.

- Get rid of the risk (eliminate the noise hazard).
- If you can't get rid of it, reduce it.
- Limit exposure (Job rotation, rest periods, etc.).

Article 5 also states that the noise in rest areas must be compatible with their purpose, and reminds us to take particular care with any particular sensitive risk groups such as pregnant women and young people.

Article 6 requires suitable hearing protection to be supplied by employers if the noise cannot be reduced and that if the upper action value is reached, the wearing of hearing protection is mandatory.

Article 7 of the new regulations specifies that the limit of 87dB(A) or an impact noise of 14dB(C) is realised or exceeded, the employer must immediately reduce the exposure, discover why is has been exceeded, and institute new control measures to prevent a recurrence.

Article 8 of the new regulations on 'Worker Information and Training' offers a specific and detailed reinforcement of the requirement for employers to train and educate workers. Workers are entitled to:

- Be made aware of all the risks.
- Be told of the control measures to be used.
- Have an understanding of the Action and Limit Values.
- Be aware of the results of Risk Assessments.
- Be trained in the correct use of any hearing protection provided.
- Details of health surveillance.
- Training in any safe working practices.

There will be a need to re-educate those workers who have previously been exposed to old levels and who are now still exposed to the new values. Attention must also be drawn to the high risk of noise-induced hearing loss.

Article 9 of the new regulations once again reinforces the requirement for employee consultation and participation. A workforce that is involved in the health and safety process is more likely to be fully cooperative in implementing and maintaining a hearing protection scheme.

Article 10 again reinforces the Management of Health and Safety at Work Regulations requirement to provide health surveillance to employees free of charge if there is an identifiable disease or adverse health condition. Noise-induced hearing loss certainly qualifies, so we should already be providing health surveillance. Audiometry is the appropriate health surveillance for noise risks, and the directive clarifies when this is a requirement. It will be a right for workers at the upper action value and at the lower action value where assessment indicates a risk to health. Current Health & Safety guidelines suggest health surveillance at 95dB(A) or 8 hour equivalent. Most health surveillance is in fact carried out on a regular basis at lower exposure levels. If Audiometry is not part of your existing health surveillance programme, then now may be a good time to start.

In areas where hearing protection is currently optional (over 80dB (A)), it will now become compulsory.

OLD	NEW
Assessment Period	8 hours or 1 week
Lower Action Level	Over 85 dB (A) Over 80 dB (A), 112 pa
Upper Action Level	Over 90 dB (A), 200 pa At 85 dB (A), 140 pa
Max Exposure Limit	87 dB (A), 200 pa
Audiometry	At 85 dB (A), 140 pa
Audiometry	Over 80 dB (A), 112 pa if risk

The table above summarises the changes to the existing regulations.

Perhaps one area not considered by the new regulations is that of 'in ear' monitoring systems used by musicians.

Technically, these systems will be covered by the new regulations but I am not at all certain how it will operate or if it is indeed 'reasonably practicable' to monitor and control such systems in respect of the regulations. Whatever is decided, we must remember that such systems are not suitable for everyone and may be almost impossible in certain situations.

Just because you have met the environmental noise conditions set in an Entertainment Licence, it certainly does not mean you meet the Noise at Work Regulations.

Environmental noise control is there to protect those who neighbour the event or venue: the Noise at Work Regulations are there to protect those working at the event or venue – they are very different.

28 THE ELECTRICITY AT WORK REGULATIONS 1989

The introduction to the Memorandum of Guidance on the Electricity at Work Regulations 1989 states: "A little knowledge is often sufficient to make electrical equipment function but a much higher level of knowledge and experience is usually needed to ensure safety." It is a statement well worth remembering.

The regulations set out in general terms the safety requirements for electrical systems and the safety of people working on or near electrical equipment. They also impose duties of compliance and, in most cases, this will be a duty of the occupier of the premises or the holder of an Entertainment Licence.

Briefly, the main points of the regulations are:

- All systems shall be of such construction so as to prevent, so far as is reasonably practicable, any danger.
- All work activity on or near a system, including operation, use or maintenance, shall be carried out, so far as is reasonably practicable, so as not to give rise to danger.
- Any equipment provided to protect people while they are at work on or near electrical equipment shall be suitable for use and properly maintained.
- The strength and capability of electrical equipment must not be exceeded in such a way as may give rise to any danger.
- All electrical equipment which may be exposed to mechanical damage, the effects of weather, temperature, wet, dirty or corrosive conditions, flammable or explosive dusts or gases, must be constructed or protected to prevent, so far as is reasonably practicable, any danger arising.
- All conductors which may give rise to danger (i.e. above 50V A.C. or 120V D.C.) must be insulated, protected, placed, and other precautions taken to prevent danger.
- Earthing or other protective measures must be taken to prevent conductors, other than circuit cables, from becoming live.
- Regulation 7 prohibits the placing of switches in the neutral side of a

circuit unless specific safety precautions are taken. I can't think of any situations in our industry where the placing of switches in the neutral side of a circuit is required.

- All joints and connections must be electrically and mechanically suitable for use.
- The use of fuses and circuit breakers to protect electrical systems is required.
- Suitable means must exist for cutting off the supply to any item of electrical equipment and isolating any such item.
- Where appropriate, circuits must be identified. (In fact, the IEE Wiring Regulations require all circuits to be identifiable).
- Adequate precautions must be taken to prevent any danger when work is taking place near equipment that has been made electrically dead, especially to prevent it from becoming live again, e.g. by the use of a 'lock out' system.
- No work must be done on or near live conductors unless it is unreasonable to make it dead, or it is reasonable to allow 'live work' and suitable precautions are taken. (The definition of live work includes the testing of a mains supply with a multi-meter. Even out of a 13A socket, if you are not competent to the 'Live Working' standards and complying with all safety requirements, you are breaking the law! The IEE Wiring Regulations in fact state that only 'competent persons' should carry out any electrical work).
- Adequate access, lighting and working space must be provided as necessary to prevent injury.
- No person to be engaged in work where technical knowledge or experience is necessary to prevent danger, unless they have the knowledge and experience or are under appropriate supervision.
- Fixed installations should be installed in accordance with B.S. 7671 Requirements of Electrical Installations (also known as the Institution of Electrical Engineers (IEE) Wiring Regulations), and then inspected and tested every five years as a minimum, when an inspection/test certificate should be issued. Local Authority Entertainment Licence conditions may call for more regular testing. The current edition of the IEE regulations makes the requirement of "ascertaining and complying with the local Licensing Authority's conditions". The conditions set in most Entertainment Licences normally call for this

same standard for temporary installations, but this is often impractical or impossible to comply with. After installation, a copy of the inspection/test certificate should be given to the production manager and/or event safety adviser.

- For temporary venues, it is normally a condition of licence that all electrical equipment be installed and tested by a qualified electrician before the licence is finally granted. A set of regulations for temporary installations is now in force. These standards are a revision of BS 5550 7.5.1. and are known as BS 7909: 1998 (A Code Of Practice For The Design And Installation For Temporary Distribution Systems Delivering AC Electrical Supplies For Lighting, Technical Services And Other Entertainment Related Purposes).
- Electrical equipment must be protected by Residual Current Devices (RCDs), since a fuse on its own may be insufficient protection. With all socket outlets of 32 Amps and below that are (or may be used) outdoors or where higher power requirements are foreseen indoors, the use of industrial connections (often known as a 'Cee Form' connector) to BS 4343 should be used as opposed to the normal 13 Amp plugs and sockets.

These outlets must be protected by a 30mA RCD. In situations where 13 Amp plugs are used, then fuses *of the correct rating* for that particular appliance must be fitted. Test Certificates for RCDs should be available for inspection by interested parties if required.

- For outdoor use, hand tools such as electric drills and saws must be, for safety, of the 110V supply type which is centre-tapped to earth so that the maximum voltage to earth should not exceed 55V. This is also a requirement of the Work Equipment Regs. that state that work equipment must suitable and safe for the work intended.
- Cables on temporary installations must be laid in such a way so as not to become a trip hazard and are protected from unintentional damage if laid across traffic (both pedestrian or vehicle) routes.
- All electrical equipment and apparatus must be fully maintained, inspected and tested by a competent person on a regular basis and records kept of all tests, inspections and maintenance.
- All portable appliances (including catering equipment, backline, effects, PA, lasers, lighting, video, projection and hand tools) should undergo Portable Appliance Testing (PAT Testing) on (at least) an

annual basis. PAT testing is not required by law but is the only recognised system for testing and so may be considered as the only option. Records of testing must be kept and be available for inspection. The BBC will not allow any equipment into its studios unless PAT certificates accompany the equipment, and many employers as part of their business safety policy will not allow employees or contractors to use their own electrical equipment unless it has been tested and permission to use the equipment been given. This includes things like kettles and radios.

- Finally, employers should inform staff that they should do a visual check on all electrical equipment such as hand tools before use and warn staff not to tamper or interfere with electrical equipment unless they are trained and qualified to do so. A visual check involves:
  - Looking to see that no bare wires are visible.
  - The outer case of the equipment is not damaged or loose and all screws are in place.
  - There are no taped or other non-standard joints in the cable.
  - The cable covering is not damaged and is free from cuts and abrasions (apart from light scuffing).
  - The plug is in good condition, for example, the casing is not cracked, the pins are not bent and the key-way is not blocked with loose material.
  - The outer covering (sheath) of the cable is gripped where it enters the plug or the equipment.
  - The coloured insulation of the internal wires should not be visible.
  - There are no overheating or burn marks on the plug, cable or the equipment.
  - Testing trip devices (RCDs) to see they are working by pressing the 'test' button every month – this keeps the mechanical operation of the RCD free from sticking.

As with all work equipment, employers and the self-employed have a duty to maintain electrical equipment in a safe working order. Method statements and manufacturer's instructions should be available for all equipment and it is vital that these items are supplied with equipment on 'dry hire'.

The regulations cover all electrical equipment, even battery powered items. Electricity presents two main types of hazard: the first, as we all know, is electric shock. Even a mild electric shock to a person climbing a ladder, scaffold or structure may cause the climber to fall. ***High voltage shocks will cause the heart to stop and can cause severe burns to the skin and body***.

At any voltage above 50 volts AC you can receive a fatal electric shock, this may be from a bare wire or another conductor, such as scaffolding or truss made live by a fault.

The second danger is fire caused by overloading a cable or other equipment (such as a stalled chain hoist motor) or by a poor connection such as loose terminal screws, dirty plugs or a plug that has not been pushed in properly.

As we know, electrical systems should always be installed, tested, checked and 'signed off' by trained and qualified electricians. All safety systems must be in place, such as Earth Leakage Circuit Breakers (ELCBs), Residual Current Devices (RCDs) and Mini Circuit Breakers (MCBs), which are better known as 'trips' and come in two types. RCDs and ELCBs provide earth fault (shock) protection and MCBs and fuses provide overload and short circuit protection.

These devices, together with isolation, fuses, correct earthing systems, cable insulation and equipment enclosures (don't leave covers off) and the placing of systems out of reach (such as overhead cables), form the basis of our electrical safety systems. Even with these systems in place it is always better to play safe and not take risks.

Employees should also be advised of the following:

- If a piece of equipment keeps 'tripping out' an RCD or MCB (more than two or three times), then that equipment should not be used until it has been checked and tested by a competent person and any fault corrected. Apart from checking that a fuse of the correct rating is fitted, there is little more a non-qualified person can do. Never try to remove or short-circuit the trip, since it is almost certainly your appliance that is at fault.
- An MCB (overload trip) is far less sensitive than an RCD (earth leakage trip). An appliance well within the rating of an MCB may well 'trip out' if it has a fault, when connected to a system with an RCD under earth leakage conditions.
- The fault may not have shown up before if it had been used on an MCB system, such as domestic installations that are not usually fitted with RCD protection.
- Any coiled mains lead will heat up in use, just like the element on an

electric cooker. To prevent this, extension leads or reels should be fully unwound from their drums before use or they may heat up, melt together and cause a fire. If a cable has to be wound, by a dimmer rack for example, it should be in figure-of-eight coils.

- Drivers of high-sided vehicles, stage builders/scaffolders, forklift drivers and other plant and machinery operators, must be made aware and on the look-out for overhead cables and the appropriate action taken. Likewise, digger drivers must be informed of underground cables and other services before starting to dig. In some cases, a Permit To Work is required before digging can commence, but this obviously only applies to open air and green field sites.
- Where generators are in use, the only safety protection on the mains leads from the generator to the distribution box/switch gear is the insulated covering on the cable. Truck drivers should be made aware of this and not drive over unprotected leads and cables on outdoor sites.
- Multi-way adapters that allow more than one appliance to be run from one socket are a major danger. Caterers have a nasty habit of using these and then wondering why the power has gone off. The reason often is that they have overloaded the circuit and the 'trip switch' has cut the power off. The rule is: one appliance to one socket.
- Production stages use many different voltages and frequencies (British, American, European etc.) which may use the same plugs and sockets. Therefore, crew should *never* assume it's okay to plug something in. If you don't *know* then ask first! Someone may be working inside that piece of equipment you are about to plug in and make it live!
- Employees should report all faults and damage immediately and that piece of equipment taken out of service until it has been repaired. Employees must be on the lookout for possible dangers such as crushed cables, damaged/faulty plugs and equipment, frayed cable, loose connections and poorly laid cables. Cables should be buried, taped, covered over or laid in such a way so as not to be a hazard.
- It is important that all connections are safe and tamper-proof, particularly in public areas.
- Connection blocks such as 'chocolate block' connectors are only

for use inside electrical housings, such as an amplifier or lighting desk. They are not designed to join up two leads or repair an extension or other leads.

- A lot of damage is done to leads simply because people do not know the correct way to coil them. They should never be knotted or coiled between hand and elbow; instead, the correct way is to coil them following the natural 'lay' and fasten them with a piece of PVC tape velcro, or reusable cable tie – not gaffa tape. This way it will uncoil naturally and you won't be putting strain on sockets, plugs and connections or doing internal damage to the lead or its covering.
- 'Lampies' should be warned that lamps on a dimmer system will still give an electric shock even with the dimmer 'off'. The System must be unplugged before doing any work such as changing bulbs. It is for this reason that only experienced technicians should focus luminaires! Parcan lanterns also have their problems. These are normally associated with damaged, broken or ill-fitting ceramics in the rear of the unit that can give rise to electric shocks. Two types of safety units are now available that operate in one of two ways: one consists of a moulded double insulated ceramic; the other has the control at the rear of the unit that prevents hands and fingers from coming into contact with live components. When buying new lamps or when replacing ceramics and other parts, these new safety units should always be purchased and used.
- PA and backline technicians should never try to get rid of hums or noise on equipment by removing the earth. Get a qualified electrician to check out the equipment and rectify the problem. The only time 'lifting' is permissible is on signal cables.
- Water and electricity don't mix! Water and most other liquids are some of the best conductors available. It is therefore essential they are kept apart. Stage managers have a duty to ensure that drinks are kept and consumed in a safe area of the stage away from any electrics, and that half-full cans and bottles are not left around by bands and crew, waiting to be knocked over onto electrical cables and high-voltage equipment with the obvious consequences.
- Mention must be made of single pole connectors such as Camloks. They are usually used as connectors rated over 100A and there are several variations available such as Litton Powerlock, BAC 2 and

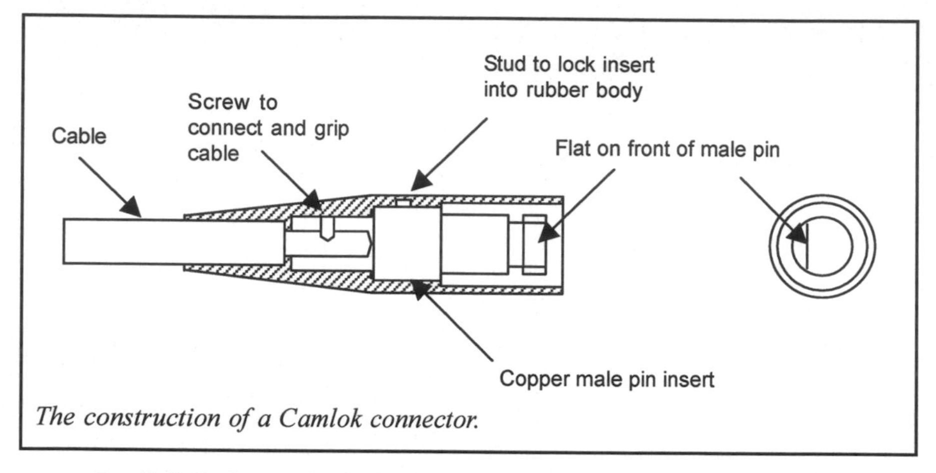

The construction of a Camlok connector.

Posilok (a form of Camlok), but they all share common principles. These are industry standard connectors for mains distribution systems and many lighting dimmer systems are fitted with this type of connection. A typical application for these connectors is a three-phase power feed or link, and the rubber outer covers are usually colour-coded to show if it's phase 1, 2 or 3; neutral or earth. They must never be used where the general public or unauthorised persons can gain access to them.
No one should touch such connections unless they are qualified or have been given clear instruction about connecting and disconnecting and they are 100% certain that that the system is 'dead' and therefore safe. Live parts can be exposed on these connectors when not mated so should never be handled 'live'. Because of this, these connectors are somewhat controversial as they allow the earth to be broken in a system without the use of tools and it is possible to touch the live connections.

The following sequence for connection and disconnection must always be followed when using single pole connectors and the power supply should always be turned off.

Connection – after ensuring the power is off, connect the female connectors first in the following order:

1 Earth
2 Neutral
3 Phase 1

4 Phase 2
5 Phase 3
6 Carry out a visual check to see that all connections are correct.

Disconnection – after ensuring the power is off disconnect the male connectors first in the following order:

1 Phase 1
2 Phase 2
3 Phase 3
4 Neutral
5 Earth

- If you require equipment to be connected to the mains by wiring in with 'bare tails', a competent person should always do this. Hire companies may like to take note of this. The best way around this is to make it a 'condition of hire' that the 'client' supplies or arranges for a 'competent' person (such as the 'house' electrician) to connect and disconnect equipment for the mains supply.

I have already mentioned 'competent' and 'qualified' persons to work on electrical installations – the Electricity at Work Regulations only refer to 'competent persons'; a person may be competent without being qualified! A 'qualified' person will be the holder of a City and Guilds 236 qualification as an electrician. The criteria for this qualification are based upon the current edition of the Institute of Electrical Engineers Wiring Regulations that also make up British Standard 7671. It is worth noting that some local authorities will only accept 'sign off' certificates from electrical contractors who are registered NIC-EIC members or who are registered with the Joint Industry Board for the Electrical Contracting Industry.

The definition of a 'competent person' in the Electricity at Work Regulations (Regulation 16) is: "A person who possesses such knowledge or experience to prevent danger or injury to himself or others having regard to the nature of the work."

I would personally add that a 'competent person' also knows his own limitations.

The subjects of qualification and competence have been the subject of many debates and will probably remain so for some considerable time.

The requirements for the colour coding of three phase and neutral conductors in electrical distribution systems are changing to harmonise the UK with the EC colour scheme. Connectors to the new colour scheme are already in

OLD	Earth (CPC)		Neutral	Line 1	Line 2	Line 3
	Yellow	Green	Black	Red	Yellow	Blue
NEW	Earth (CPC)		Neutral	Line 1	Line 2	Line 3
	Yellow	Green	Blue	Brown	Black	Grey

Colour Code Changes

circulation within the industry and the Amendment to BS7671 (the 'Wiring Regs.') became effective on 1st April 2004.

Use of *either* the old colour coding *or* the new colour coding (but not both) will be permitted on new installations until 1 April 2006. After that date, the 'new' colour coding must be used on all new installations. The introduction of changed colour codes will apply to temporary electrical installations used extensively in the live music industry at shows and events.

Those responsible for supply, hire, planning, installation and use of temporary electrical installations and equipment, need to be aware of the impact of the changes and the possible safety implications to avoid potential electrical and safety hazards. This is particularly important where single pole connectors may be in use and where a mistake in connecting due unclear colour coding may quite possibly be fatal.

The view is taken that a temporary distribution system as used in the entertainment industry at a venue or on site, etc. would be considered a ***new installation***.

The Electricity at Work Regulations state: "No work must be done on or near live conductors unless it is unreasonable to make it dead, or it is reasonable to allow live work and suitable precautions are taken".

Live work (also known as 'hot work') includes the testing of supplies to see that they are live, that the phases are correct and the supply is clean and stable. Obviously we can't carry out these checks unless the system is live, so we must take *every* possible precaution to make the operation safe. This will include using G57 fused leads on our test meters and testing the meter before we start to show that our readings are correct; we don't want it to show dead when the system is really live. The control systems to be used during hot work include the use of barriers to prevent unauthorised access into the work area, the provision of adequate lighting, and having an observer present.

A more recent development in electrical safety involves the use of a market-leading brand of chain hoist motors as used in rigging operations. Until September 2002 these motors were supplied with 3-pin, 110V, 16 Amp CEE form connectors. Since September 2002 all new chain hoist motors are now fitted with a 4-pin, 110V, 16 Amp CEE form connector, this change has been agreed by the HSE after long negotiations with the HSE and various trade associations.

The reason why the 3-pin, 110V, 16 Amp CEE form connectors were not suitable was because the earth pin on the connector was used as a conductor and this precluded its future use.

The HSE noted in its statement that, although both EN 60204-1 and 60204-32 preclude the use of EN 60309-2 CEE form plug and socket connectors in control applications (as they are mains connectors), there may be justifiable reasons for using such connectors, providing that the earth pin is not used for any purpose other than as a protective conductor.

The technical reasons for such selection should be contained in a technical file, together with the deviations from the standard stated in the Certificate of Conformity supplied with the product. All existing motors fitted with a 3-pin connector should be changed to the 4-pin type. Full details can (and should) be provided by the supplier.

29 THE WORKING TIME REGULATIONS 1998

Good news for the genuine self-employed – these regulations don't apply to you! This is all well and good but the HSE is looking at long working hours as a hazard and will still want risk assessments from self-employed persons working over 48 hours a week. Just because you are self-employed does not mean you can (legally) work until you drop.

Until these regulations came into force, with the exception of most goods and passenger vehicle drivers, the law had not set specific limits on the number of hours people over the school-leaving age can work, but all that now changes. The regulations were set by the Department of Trade and Industry (DTI), but enforcement will fall into the hands of the HSE. The main problem with these regulations is that they are very complicated and difficult to understand. I am sure that a higher level of compliance would be attained if they were simplified and more 'user friendly'.

All employers have a general duty under Section 2 of the HASAWA to ensure, so far as is reasonably practicable, the health, safety and welfare at work of all their employees. This means they cannot require people to work excessive hours or unsuitable shift patterns likely to lead to ill health or accidents caused by fatigue. Work schedules should also allow for adequate rest periods. That includes staff driving up and down the motorways to attend the endless string of meetings we need to attend as well as crew travelling between jobs. Fatigue increases the risk of accidents and accidents are going to result in reduced productivity.

Under Regulation 3 of the Management of Health and Safety at Work Regulations 1999, employers are required to carry out risk assessment to identify hazards such as fatigue, and evaluate the extent of the risks involved, so that measures can be taken to comply with the general duties under the HASAWA.

It was on 1st October 1998 that the Working Time Regulations came into force to reinforce the above. When considering the main points of these regulations remember Health and Safety considerations are paramount and employers must ensure that proper written records are kept of hours worked per week, night workers, breaks, compensatory rest periods, work patterns and annual leave.

Excluded Sectors

There are a number of categories of people who are excluded from the regulations relating to maximum working week, maximum hours of night work, health assessment provision, rest breaks and annual leave provisions. Everyone is our industry is affected by these regulations; we are not one of the excluded sectors.

In 2002 the government announced an extension to the regulations to include employees previously excluded from the regulations to include:

- All railway workers such as train drivers and ticket inspectors.
- All non-mobile and certain mobile workers in road, sea and inland waterway transport and sea fishing – for example, clerical and administrative staff.
- Offshore employees – such as oil rig workers.
- Junior doctors, and
- Workers in the aviation industry that are not covered by the Aviation Directive – such as baggage handlers.

48-hour Week

All workers under a contract of employment (including agency workers) are, from 1st October 1998, subject to a maximum working week not exceeding 48 hours.

As mentioned, certain workers are excluded from this provision, but not us!

All other employees may modify the effect of this regulation by devising a collective agreement with unions or a workforce agreement extending the reference period from 17 weeks to 52 weeks. The Reference period is the time span over which the hourly week is calculated and unless modified must be 17 weeks.

However, workers and employers can enter into individual voluntary agreements to dis-apply the 48-hour maximum. These agreements must specify the number of hours the worker is agreeing to work.

Vague descriptions such as 'when required' will not suffice. Agreements must be in writing and should have a notice of termination clause not exceeding 3 months. Where such an agreement is in place, employers must keep a record of the number of hours worked. Being 'on call' at home will not generally constitute working time.

At the time of writing the European Commission was considering an end to the existing 'opt out' from the 48 hour working week after Members of the European Parliament (MEPs) voted in favour of the opt out being gradually phased out.

Annual Leave

All workers are now entitled to 4 weeks' paid annual leave, including bank holidays. Part-timers will receive an equivalent entitlement, which will be calculated on a pro-rata basis. Employers should ensure their contracts of employment reflect this and contain notice provisions for the taking of annual leave. There is an implied duty on employers to ensure employees take their leave entitlement. There is no right to carry over holiday or to pay wages in lieu (unless the employee is leaving).

On leaving, the employee will be apportioned days on a pro-rata basis and be paid in lieu. The leave year can be determined by written agreement; otherwise it will start on the 1st October.

The regulations specify (in the absence of any agreement to the contrary) the employee must give the employer notice to take leave which is double the amount of the holiday taken, e.g. 2 weeks' holiday = 4 weeks' notice. The employer only has to give notice equivalent to the holiday taken, e.g. 1 week's holiday = 1 week's notice. As these statutory notice provisions are complicated, I advise all employers to have specific notice leave clauses in their contracts of employment, i.e. specify how much notice is required from either employer or employee of any request for holiday or refusal of such a request.

Employers can include a written claw-back clause for holiday taken before it has been earned.

Rest Breaks

Adult employees (over 18 years) are entitled to 11 hours' rest in every 24-hour working period, not less than 24 hours in each 7-day period or 48 hours in each 14-day period.

The employer may decide which of these to apply. Employees are also entitled to 20 minutes' rest after working for 6 hours. These rest breaks must be uninterrupted.

Young Workers (under 18 years of age) are entitled to 12 hours' rest in any

24-hour working period, 2 consecutive days, starting from midnight in each 7-day working period and 30 minutes' break after working 4.5 hours. Breaks are unpaid and can be extended by agreement but not excluded. Where work is of a monotonous nature or hazardous to health, the employer must ensure adequate breaks are taken.

Night workers

The regulations state maximum hours. Any employee who works at least 3 hours of their daily working time during the hours of 11pm and 6am is a night worker. Workforce, collective or individual agreements may be drafted to include a night period different from that specified, but the period must be of at least 7 hours which must include midnight to 5am.

Employers are required to take all reasonable steps to ensure that night workers (unless they fall within an exempt category or derogation) do not work more than one 8-hour period in any 24-hour period. Using a reference period of 17 weeks to calculate the number of hours worked, the period starts from the date specified in a workforce, collective or individual agreement or the date the employee started employment if after 1st October. In practical terms, Health & Safety bodies can use a snapshot of any 17-week period during the working year to calculate hours worked. All night workers are entitled to a free health assessment before starting night work and at regular intervals thereafter. This can be done by the completion of an initial health questionnaire then referral to a doctor if necessary. A night worker is entitled to be transferred to day work on request if they have medical evidence indicating that night working is affecting their health. They are not entitled to transfer with any enhancements such as extra pay.

The employer is not required to dismiss a day worker to make room for such a transfer but is required to reassign workers, if possible, subject to their agreement. Employers must not contract out of the right to health assessment or transfer.

Derogations/Exemptions

Unmeasured Working Time

This covers workers whose time cannot be measured or predetermined and covers managing executives, family workers, and workers officiating at religious ceremonies in churches and religious communities. Such workers are not subject to regulations governing a 48-hour week, length of night work or rest breaks.

Work Specific Activities

Includes security and surveillance workers and 24-hour service providers, e.g. dockers, medical workers, hotel staff, postal workers, farmers, tourism workers, refuse collectors, gas, water and electricity providers.

Regulations relating to a maximum working week, night work and breaks do not apply.

Shift Workers

These are workers whose work schedule is part of shift work, whereby each worker replaces another at the same continuous work station e.g. factory machine workers. Such workers are exempt from the rest break provisions. NB. Although exempt, these workers are entitled to compensatory rest, which constitutes a rest period equivalent to that specified in the Regulations to be taken at times other than those specified in the Regulations. The employer must observe the entitlement to compensatory rest.

The above derogations are subject to any relevant Health and Safety Regulations, Health and Safety being the prime consideration at all times.

Enforcement

Tribunals will enforce rest breaks, annual leave and compensatory rest. Claims must be brought within 3 months of the act complained of and, if successful, awards of compensation may be made. The employer must not subject employees who make a complaint to an employment tribunal to any detriment; such treatment would give rise to a claim.

Health & Safety bodies will enforce the maximum working week, maximum night work, and duty to provide health assessment. Sanctions include criminal penalties such as conviction, fine and imprisonment. Health & Safety bodies have the right of access to information and records. If an employer fails to comply, criminal sanctions are available. Employers must keep proper written records of night workers, all hours worked, compensatory rest periods, work patterns and annual leave.

Workforce Agreements

These Agreements must be in writing and specify the start date and only apply to relevant members of the workforce. To be valid, they must be signed prior to the start date by the majority of those who are relevant members of the workforce or by elected representatives. These representatives must be elected by secret ballot.

The employer is free to determine the number of representatives to be elected and the term for which they are elected. All members of the workforce are eligible to stand as candidates and all of them are entitled to vote. To be valid, the majority of the workforce must agree and have seen copies of the agreement prior to signature.

Individual Agreements

These are agreements entered into voluntarily by the employee and employer. They must be in writing and cannot be used to exclude regulations in their entirety but may be used to modify the effect of the regulations.

Collective Agreements

These are agreements reached between an employer and a trade union recognised by that employer for collective bargaining purposes.

They do not have to be in writing, but to be enforceable they must be incorporated into the contract of employment, either by reference or expressly.

What record keeping obligations do the regulations lay down?

- There is a general obligation for all employers to keep records adequate to show compliance with the maximum working week and the night work provisions relating to average and maximum daily hours of work and health assessments.
- There is a specific obligation for employers to keep records of the working time of those workers who have opted out of the 48-hour limit.
- Both these obligations are enforced as Health & Safety offences.
- What are the other obligations imposed by the regulations?
- Not to assign a worker to night work without offering him the opportunity of a free health check (in the case of young workers, a check on health and capacities) at the outset and at regular intervals while he is a night worker.
- Where possible, to transfer a night worker suffering health problems, which a doctor has advised are connected with night work, to day work to which the worker is suited.
- To ensure adequate rest breaks where the pattern of work (e.g. monotonous work or work at a pre-determined work-rate) puts

workers' health and safety at risk.

All of these obligations are enforced as Health & Safety offences.

I'm sure we're going to hear a lot more about the Working Time Regulations – there is a lot more to it than I have had time to cover here and a lot of questions still remain unanswered. I was once asked to advise a tour manager who was in dispute with a well-known artist. The artist in question insisted that the crew for the artists tour travel in a minibus that only allowed a couple of hours each night in the hotel and therefore only allowed a couple of hours' sleep each night.

The tour manager had budgeted for a sleeper bus but the artist would not allow for this, even though it was proven that the sleeper bus worked out cheaper than a minibus and hotels. The risk assessments and the information the tour manager was provided with were major factors in persuading the artist to make an out of court settlement for unfair dismissal!

With luck, these new regulations may end a recent trend that has involved promoters and bands 'playing off' service companies against each to obtain cheap quotes for low budget tours. These quotes can only be sustained by expecting crew (such as sound engineers) to double as goods' vehicle drivers *in addition* to their (sound engineering) duties for the tour.

This technique has been applied by some promoters during 'slack' work periods when they know the service companies are in need of work.

It is not only immoral for the promoters and bands to use such tactics, it is totally illegal for the drivers and operators for whom they work. They both risk major prosecution, and rightly so. The maximum number of hours a driver can drive per day in the UK is ten a day with a maximum of eleven hours on duty. In Europe, the number of hours is reduced to nine, but this is only a 'rule of thumb' guide because the regulations in Europe are much more complicated.

This includes *all* work, not just driving. We should ask if truck drivers should also be working as follow spot operators. In fact, are the truck drivers trained, equipped and insured to climb into the truss to operate follow spots? Probably not!

A change in the law will now make it possible for employers to be prosecuted should an employee have a traffic accident (even after work on the way home) that can be attributed to the employee being overtired due to excessive working hours. I can think of at least one friend who has been involved in very serious accidents when they feel asleep at the wheel while driving home from a series of gigs held over several days during which time he had little or no sleep as he was working as a driver as well as sound engineer.

Employers have a duty of care as well as a legal duty to ensure that employees don't drive for long hours after carrying out other duties. Should an accident take place, all a court has to do to obtain a prosecution is prove three points:

1. That there was a duty of care. (Very easy to establish as there is always a duty of care).
2. Prove that there was a breach of that duty of care.
3. Prove that the accident was as a result of that duty of care.

It is known that the HSE is aware of how some companies operate and they have already prosecuted a transport company whose driver fell asleep at the wheel and caused an accident! So it's not just the driver who gets nicked but his employer as well and it does not have to be a company vehicle, it could be the driver's own private car he was using to get to and from jobs or to get home after work.

Excessive work hours also lead to stress and the HSE is currently putting a great deal of emphasis on stress and stress-related problems. Employers have a duty to protect employees from stress so risk assessments must include this often-overlooked condition.

Recently it has been suggested by medical researchers that irregular eating habits and long stressful work hours, particularly at night, can lead to diabetes. Several friends (including BBC Presenter John Peel) and I, who have worked in our industry for many years, have all recently been diagnosed as being diabetic despite the fact there has been no family history of the disease. With the number of people now being affected by diabetes this is more than a coincidence and is obviously an area that requires further research.

In the summer of 2004, the Department of Transport (DfT) announced proposed new changes to the number of hours a truck driver can work.

The Road Transport Directive (RTD) will be implemented into UK law in March 2005 and will affect drivers of HGV and public sector vehicles (such as buses) whose work is covered by European Drivers' Hours Rules.

Under the RTD, drivers will have the right to an average maximum working week of 48 hours with up to 60 hours of work allowed in a single week. However, unlike other UK working time legislation, the Directive will not allow individuals to opt out of the weekly limit. Drivers average weekly working hours will be calculated over a four-month reference period, which can be extended to six months if there is a collective agreement in place between an employer and their workers.

In addition, the RTD will give drivers the right to a maximum limit of ten hours work in each 24 hour period if night work is undertaken. Nightwork is defined as the four our period between midnight and 4 am for HGV drivers and 1 am and 5 am for the drivers of passenger vehicles. However, drivers will be allowed to drive for longer hours if a collective agreement is drawn up.

Under the new regulations, drivers will not be allowed to work more than six consecutive hours without taking a break and that at least 30 minutes will be required if working between six and nine hours a day while 45 minutes must be taken if working hours exceed nine hours a day. These breaks can be spilt into 15 minute periods.

The changes do not apply to self-employed drivers but we have to be very careful about what is a self-employed driver, the definition of a self-employed driver is an 'owner driver'.

The new regulations will be enforced by Vehicle Operators Services Agency (VOSA), a government body formed to improve road safety and environmental standards.

Action To Take

Do any workers work over an average of 48 hours per week?

No - Consider how you can demonstrate compliance if asked by the Health & Safety enforcing authorities. Can existing records prove this? Would a monitoring system need to be put in place?	Yes - Does the worker want to sign an agreement to work over this limit?
	NO - Then rearrange work patterns so that this worker is not forced to work over the weekly working time limit.AndConsider how you can demonstrate compliance if asked by the Health & Safety enforcing authorities. Can existing records prove this? Would a monitoring system need to be put in place?
	YES OPT-OUT Make an agreement in writing and keep records of the hours that the worker works.

30 SLIPS, TRIPS AND FALLS

Generally speaking, slip and trip hazards are covered by the Workplace (Health, Safety and Welfare) Regulations 1992.

The following items can cause trip or slip hazards:

Ramps
Cables
Uneven ground
Items of equipment and flight cases
Edges of stages and platforms
Risers
Guy ropes and pegs
Tent door flaps
Temporary roadway
Spillages
Mud, Rain, Ice and Snow.

Falls, on the other hand, come under four categories:

Falls from the same height
Falls into drops or holes
Low-level falls from ladders
Climbing falls

(See the section on 'Work At Height Regulations 2004' for information on low level and climbing falls)

Perhaps with the exception of falls, most of these hazards can be removed with good 'housekeeping'. Holes and trenches should be filled in as soon as possible and protected while they are open by covering, fences, marker tape, stakes and pins. Warning signs should also be erected. Cables must be laid in such a way so as not to be a trip hazard; they may also be taped down or covered with a cable ramp or piece of carpet that is well secured by gaffa tape or similar.

Handrails should be provided around stages, platforms, steps and ramps, a lower rail should be provided to prevent flight cases rolling underneath the top rail and falling, thus damaging equipment and anyone who may be unfortunate to be underneath.

Spillages (especially chemicals) must be cleared up promptly, and keep all crew drinks on a table on one side of the stage away from all electrical systems. Only allow artists to bring essential stage drinks onto the stage then appoint a crew member to clear these up at the end of the performance before risers and equipment are moved. It only takes a few seconds and could prevent major problems. Keeping the workplace clean and tidy prevents accidents of many kinds, including slips and trips.

Where else would you find on a sheer drop on a work site that has no protection, no warning sign or barrier? I am of course referring to the 'downstage' edge of a stage. Well of course most of these things would look ludicrous when put across the stage for a performance, but during a load in or load out it has become common practice in theatres in the UK to put a barrier across the stage to prevent people falling into the pit. There is no reason why we can't use similar safe practices in our industry. All open edges (usually the downstage edges) of stages, platforms, risers and ramps should be clearly marked during a performance with a strip of white paint or gaffa tape about 45 mm wide; claims have been made by artists who have fallen off stages that did not have the edges clearly marked. Another idea (during a performance) is to put small LED lights across the downstage edge to remind those onstage where the edge is.

Tent pegs can be protected with pieces of foam rubber. The foam rubber tubing used to insulate pipes is very good, or even polystyrene cups held on with white tape and a dab of white emulsion paint will highlight guy ropes.

Entertainment Licence fire safety conditions often require entrances and exists of tents and marquees to be proper doors that open outwards and are mounted in a frame. Flaps are generally not allowed and any guy ropes must be kept well clear of door openings so that no hazard exists in the event of an emergency evacuation.

31 WORK AT HEIGHT REGULATIONS 2004

At present, the regulations regarding work at height come under the Construction (Health, Safety and Welfare) Regulations 1996, which set a general duty to prevent falls of people 'so far as is reasonably practicable'. The Workplace (Health, Safety and Welfare) Regulations 1992 require measures to be taken to prevent the risk of falling a distance likely to cause injury – this could be a distance as little as six inches and obviously this does not allow for fall arrest systems or work restraint techniques to be used.

Therefore, for the type of work we often undertake, we currently come under the Construction Regulations rather than the Workplace Regulations in respect of work at height. This situation is due to change in the very near future with the introduction of the Work at Height Regulations 2004. From the information I have seen so far the new regulations will just underpin existing regulations such as the Management of Health and Safety at Work Regulations, PUWER, CDM, the PPE Regulations and LOLER.

The information provided here on the new regulations has been extracted from the Statutory Instruments (the Law) relating to the regulations and from the draft guidance documents. The HSC/HSE will be publishing an Approved Code of Practice and Guidance to the regulations in due course that will undoubtedly give much more detailed information on how to comply with the regulations.

Regulation 6 (Avoidance of Risk from Work at Height) states that work at height must not be carried out unless a risk assessment has shown it cannot be carried out more safely by any other means other than working at height (nothing new here, a normal risk assessment will show exactly the same). The regulation also sets out the hierarchy of measures you should follow when performing work at height. Following your risk assessment this hierarchy should allow you to select the most appropriate methods for work at height. The overriding principle is to prevent, so far as is reasonably practicable, any person falling a distance liable to cause personal injury.

1	*Avoid* the risk by not working at height - where it is reasonably practicable to carry out the work safely other than at a height to do so.
2	*Prevent* falls - where it is reasonably practicable to avoid work at height, you should assess the risks and take measures to allow the work to be dfone whilst preventing, 'so far as is reasdonably practicable', people or objects falling. This might include ensuring the work is carried out safely from an existing place of work, or choosing the right work equipment to prevent falls.
3	*Mitigate* the consequences of a fall - where the risk of people or objects still remains you should take steps to minimise the distance and consequences of sich falls. This also involves the selection and use of work equipment.
4	At all stages give collectove measures (e.g. guardrails, nets, airbags, etc.) precedence over personal protective measures (e.g. safety harnesses).

The Hierarchy for Safe Work at Height

Within this framework the Work at Height regulations require you to:

a) Assess the risk to help you decide how to work safely.
b) Follow the hierarchy for safe work at height – avoid, prevent, mitigate, and give collective measures priority.
c) Plan and organise your work properly, taking account of weather conditions and the possibility of emergencies.
d) Make sure those working at height are competent.
e) Make use of appropriate work equipment.
f) Manage the risks from working on or around fragile surfaces and falling objects.
g) Inspect and maintain the work equipment to be used and inspect the place where the work will be carried out (including access and egress).

The Regulations include requirements for:

- Work at height to be planned and appropriately supervised so that it is carried out in a safe manner [Regulation 4].
- Work tasks performed on working platforms.
- The use of ladders.
- The use of rope access and positioning techniques.
- Steps to be taken to prevent persons falling through any fragile

material [Regulation 10] (not really an issue in our industry).

- Steps to be taken to prevent the fall of any material or object during work at height [Regulation 11].
- The inspection of working platforms and personal suspension equipment. [Regulation 12] (already required by PUWER, the PPE Regulations and LOLER).

The new regulations contain five Schedules dealing with guardrails, working platforms, personal protective equipment, means of arresting falls, and ladders. The general view of most Health & Safety professionals is that these new regulations are long overdue and only go to reinforce what the good guys have been doing for a long time. It looks as though there will be nothing really new.

We must remember that it's not just riggers that work at height; others include stage and set builders, electricians, suppliers of drapes and décor, pyrotechnicians, lampies, laser technicians and on occasions, even PA crew.

Each year an average of almost 14,000 people are reported as receiving injuries or having been killed after falling from height. During 1998/99 statistics indicate that 8809 people were injured, including 10 fatally, from falling from a distance of 2m or less. The HSE put work at height on its high priority list for needing attention. There are many reasons why people fall but the main factor is something called gravity that pulls us towards the ground at an alarming rate. Unfortunately, there seems to be nothing we can do in practical terms to get rid of this gravity stuff, otherwise we would be able to reduce this accident rate to zero.

If at all possible, the requirement to work at height should be avoided at all costs. Following two fatal accidents, Earl's Court arena in London installed a grid system that can be lowered to ground level. Now a touring production can install all rigging points on this grid, hang trusses, lamps and all associated equipment without the need to work at height – an ideal situation and one that is already in use in many major venues, but sadly not all.

Prior to the new regulations, work at height was considered to be anything over two metres from the soles of the feet to the floor, hence for any place used as access or to work from where there is a risk of falling 2 metres or more. But, the new regulations removed the 'two metre rule' as so many serious accidents take place involving falls of less than two metres. Therefore, decisions as to the working techniques have to be taken following the hierarchy of selection of suitable equipment in the order that follows. There are *no* exceptions to these regulations:

1) Guardrails or barriers or other 'similar means of protection' have to

be provided to prevent falls. The guardrails, etc. should comply with the standards set out in **Schedule 1** of the regulations (Requirements for Guardrails, etc).

2) If there is no safe place to walk or work, then a working platform (mobile or static) is required which complies with **Schedule 2** of the regulations (Requirements for Working Platforms). Working Platforms include scaffolding.
3) If it is not practicable (which means, possible in the light of current knowledge) to comply with all of the above requirements, or if the nature of the work is difficult, or of very short duration, then personal suspension equipment (Rope Access Equipment) should be used.
4) If again it is not practicable to comply with any of the above or the nature or duration of the work is as detailed as in the last paragraph, then a means of arresting the fall of the individual (Personal Fall Protection System) should be provided and used only if Collective Safeguards for Arresting Falls (such as airbags, nets or similar) are not suitable and practicable.

Schedule 3 gives the requirements of Collective Safeguards for Arresting Falls and **Schedule 4** details the objectives to be met in the use of Personal Fall Protection Systems (including additional requirements for Work Positioning Systems).

It must be remembered that most access to work places at height will probably take place by using a ladder rather than climbing.

Climbing and the use of Personal Protective Equipment must only be considered as a last resort, if none of the other systems can be utilised. The reason for this is that PPE is a *protective* system, whereas all the others are *preventative* systems.

The reasons why many big companies purchase access machinery is not surprising, since to reduce the risks to workers to an acceptable level is a legal requirement.

When the need for access is clear, the business will not be able to justify *not* having a cherry picker on site.

The risks can be reduced because the employer has an element of control over how work is done and who is doing it.

A variety of jobs can be undertaken and when the reduction in risk is considered, the economics of owning a machine start to look more favourable.

There are times when climbing access is necessary in order to be practical. The size of the building, quantity and locations of rigging points (for example) and available roof access can mean that a MEWP (Mobile Elevating Work Platform) is not viable. The site may not have a floor that can cope with the weight or doors that are big enough to allow access for a MEWP, and so on.

There will be times when it is necessary to access flown structures that cannot be reached by conventional means and many roof spaces are not well served with catwalks. Provided there is a safe system of work in place and a sufficient number of trained and able staff, there is no reason why work at height need be dangerous. Using a well thought out fall arrest and rescue policy will allow personnel to think about their work rather than worry about dropping things or falling.

Anyone working at height must be physically fit (in accordance with advice from the Employment Medical Advisory service) and not suffering from epilepsy, asthma, obesity, heart or back disorders.

Many Public Liability and Employers' Liability Insurance Policies exclude any work above two metres or particular types of work at height, for example rope access systems. The advice is: check you have adequate insurance if you intend doing any work at height and, if required, upgrade your insurance to give adequate cover.

Before starting work at height a risk assessment including the following points should be carried out.

The Task

The need to do the work at all – can it be done later in a different way? The need to work at height – is it important enough to involve a risk, however slight?

Should fixed, wheeled or wheeled moving access platforms, rope access or fall arrest equipment be used? Is the operation likely to involve roping items or materials up or down?

Is simply using the equipment creating more hazards and risks to the worker when a fall occurs? Can an unconscious worker be rescued within a safe period of time?

The nature, extent and time frame should be assessed before deciding which access equipment is most suitable. A fatigue factor will play a role in this assessment.

The Equipment

Is the equipment best suited for the job? Is other more suitable equipment available? Is the equipment properly maintained? Are there any broken rungs? Is the equipment properly erected?

If, for instance, outriggers have been omitted, have other precautions been taken? e.g. the use of ties or additional techniques to provide stability or by the use of safety harnesses independently and adequately supported? Is the necessary safety equipment on site? Do the people doing the work have the appropriate training to use it? Has it been inspected and is it safe to use in the circumstances that exist on site?

In the regulations, **Regulation** 7 requires the following points to be taken into account when selecting suitable work equipment:

1)
   a) The working conditions and the risks to the safety of persons in the place where the equipment is to be used.
   b) In the case of work equipment for access and egress, the distance to be negotiated.
   c) The distance and consequences of a potential fall.
   d) The duration and frequency of use.
   e) The need for easy and timely evacuation and rescue in an emergency.
   f) The later provision of the new regulations.
   g) Give collective protection measures priority over personal protection measures.

2) An employer shall select work equipment for work at height which:
   a) Has characteristics including dimensions which –
      i) Are appropriate to the nature of the work to be performed and the foreseeable loadings
      ii) Allow passage without risk.
   b) Is in other respects the most suitable work equipment.

Selecting equipment for access or egress will depend on the particular use envisaged. For frequent access, you should consider more permanent arrangements. For example, if a scaffold is to be in place for some time, the erection of a staircase with handrails would be more appropriate than a ladder tied in place, especially if bulky loads are being carried up a long flight. You should also consider the use of hoists or other methods if this will reduce the risk of falls.

The WAHR recognise that work at height can be performed safely in a number of different ways, using a wide range of work equipment. The choice of equipment will depend on the risk assessment – different types of equipment will have advantages and disadvantages depending on the task and the environment in which the work is to be performed. Whatever equipment is selected it should be of sound construction in suitable material, be of adequate strength and be free from obvious defects. It must also meet any specific requirements set out in the WAHR Schedules.

Regulation 8 (Requirements for Particular Work Equipment) requires **Schedules 1 –5** of the regulations to be complied with in each particular case.

Regulation 9 (Continuity of Means of Protection) requires **Schedule 1** to be complied with at all times except at a point where there is a ladder or stairway or for the purposes of access or egress. If these safeguards are removed, then work must not take place until effective compensatory measures are in place.

The Worker

Is the technician able to function confidently at height and on the proposed access equipment? Consider the task, the time scale, the environment, climate and workloads. Is the technician properly trained and experienced?

Can you prove the technician has the skills, training and experience needed to gain access safely, including getting back down again and what to do if things go wrong?

Regulation 5 requires every person engaged in work at height (including organisation, planning and supervision) to be 'competent' in relation to the work at height or work equipment for use in such work unless he or she is being trained or being supervised by a 'competent person'.

The draft guidance to the new regulations gives one of only two known definitions of 'competence' contained within Health & Safety regulations as follows:

Competence is a combination of appropriate practical and theoretical knowledge, training and experience, which collectively should enable a person to:

- Undertake safely their specified activity at their level of responsibility.
- Understand fully any potential risks related to the work activity (tasks

and equipment) in which they are engaged, and

- Detect any defects or omissions and recognise any implications for health and safety with the aim of specifying appropriate remedial actions that may be required in relation to their particular work activity. This could include refusing to do a particular task if the potential risk is assessed as being too great.
- Undertake safely their specified activity at their level of responsibility.

It is not known if this (or any definition) of competence will be included in the guidance that is finally produced. The only other known definition of 'competence' is contained within the Electricity at Work Regulations.

Regulation 14 (Duties of Persons at Work) requires all employees or those under the direction or control of another person to report any activity or defect relating to the work at height which he knows is likely to endanger the safety of himself or another person. Every person working at height must also use any work equipment or safety devices provided in accordance with any training received and any instructions which have been provided in compliance with the requirements and prohibitions imposed or under relevant statutory provisions.

Employees using their own equipment for work at height: the WAHR applies to personally owned equipment used for work at height. Employers need to ensure that such equipment is checked and assessed as being suitable. This is particularly important where an employee brings his own tools onto the site and where the employee chooses to use his own equipment for work at height (e.g. safety harnesses).

The employer needs to establish who will be using such equipment (especially where it might be shared) and that the users are clear as to how to use it. The employer should also ensure that safe loading is adhered to and that is compatible with other safety equipment, such as anchor points. An employer should ensure that any personally owned tools are appropriate for the task, are in good condition and can comply with the Health & Safety management controls identified in any risk assessment.

The responsibility for the safe application and use of personally owned tools and equipment cannot be derogated to those people carrying out the work.

The Location

Is the equipment placed so that the technician can readily use both hands to do the task required? Are lighting conditions adequate to permit safe working?

Do not move temporary access equipment if there is insufficient light to do so safely. Do you need to barrier off the area beneath?

Falling Objects

Regulation 11 requires suitable precautions to be taken to prevent falling objects or materials causing injury. Have precautions been taken to prevent risks of injury from falling tools or equipment? Wherever possible tools should be attached to the technician or the access equipment. The technician should empty his pockets before working at height: a 50p piece falling from 30ft can kill a person below! Work equipment such as buckets or colour frames should be sited carefully so as to reduce the risk of dislodgement. The use of kickboards to contain equipment on catwalks, access towers and scaffolds may well be a controversial requirement.

Training

It is essential that those erecting/installing equipment and working at height are fully trained in the procedures as well as being supervised by a trained person.

With regard to fall arrest and work positioning, it is important to realise that it is not sufficient for workers simply to wear harnesses. They need to be trained and disciplined to use the equipment correctly.

When purchasing equipment, consideration must be given to ensuring that the equipment will in fact afford the greatest degree of protection to the user. Such issues as equipment compatibility, comfort, ease of fitting, adjustment, use and training must all be considered.

Training must also include the selection of suitable anchor points and must be effective to ensure that the user understands why the equipment is to be used, how to effectively attach and use the system and its component parts, plus the ability to inspect and care for the equipment so that it may assist in preventing a fall or protecting the user from injury.

The roof of a venue is not the place to learn how to abseil to an injured colleague.

Educating those with a need to work at height is vital.

Training is the only way to ensure any measure of competence and the first thing to appreciate is just what *can* go wrong.

Inspection of Work Equipment

Regulation 12 of the Work at Height Regulations requires almost all equipment used for work at height (including work platforms both mobile and static, scaffolds, ladders, work positioning, rope access, work restraint, and fall arrest equipment) to be inspected by a 'competent person' at suitable intervals, and each time the equipment is subjected to circumstances that could jeopardise the safety of the equipment. Some work equipment such as MEWPS and rope access equipment are already covered by LOLER in terms of inspection requirements and there will be no need to double up on inspections for equipment already covered by LOLER.

All work platforms (except man platforms fitted to lift trucks and thus covered by LOLER), scaffolding, tallescopes, work positioning, work restraint, fall arrest systems and ladders will now need to be inspected by a 'competent person' and the results recorded.

The regulations require that no work equipment leaves an employer's undertaking or, if obtained from the undertaking of another person, is used in his undertaking unless it is accompanied by physical evidence that the last inspection required to be carried out by this regulation has been carried out.

Inspection means such visual or more rigorous inspection by a 'competent person' as is appropriate for safety purposes and includes any testing appropriate for the purposes.

Employers must ensure that the results of inspections are recorded and copies kept until the next inspection under this regulation is required.

Inspection records must be kept at the employer's office or if the work at height is not at premises permanently occupied by him, at the site of the work equipment until work at height on the site is completed and thereafter at his office for a period of 3 months after completion of the work at height. Extracts and copies of inspection records must be sent to HSE Inspectors if requested.

Equipment that has to be installed or assembled must also be inspected after installation or assembly before it is put to use. A work platform other than a MEWP, a work platform from which a person cannot fall more than 2 metres or a fall arrest system must not be used in any position unless it has been inspected in that position within the last 7 days.

A person carrying out inspections must complete a written report containing the particulars set in **Schedule 6** of the regulations during the working period in which the inspection was carried out and must also provide the report or a copy to the person on whose behalf the inspection was carried out within 24 hours of the inspection. This report may be sent by electronic means.

Schedule 6 (Particulars to be included in a Report of Inspection) requires the following information:

1) The name and address of the person for whom the inspection was carried out.
2) The location of the work equipment inspected.
3) A description of the work equipment inspected.
4) The date and time of inspection.
5) Details of any matter identified that could give rise to the health and safety of any person.
6) Details of any action taken as a result of any matter identified in paragraph 5.
7) Details of any further action considered necessary.
8) The name and position of the person making the report.

Regulation 13 requires that the surface of all places of work at height be inspected visually on each occasion before use. A competent person should carry out this inspection but there is no requirement to record the inspection.

Rescue

How many people will be needed if things do go wrong?

All staff working at height using fall arrest safety harnesses and lanyards must be trained in appropriate rescue technique and operations.

To use any fall protection system within safe limits requires the formulation and provision of a rescue plan. The plan needs to account for the number of workers likely to be involved and their experience. A quick and efficient rescue needs to be ensured.

Because rescues will be few and far between, the skills required need practising on a regular basis. It is to be hoped that people will never need to use their rescue training.

Any rescue 'team' members will need special training because of the teamwork involved. This is not basic awareness training, but a specialist skill that requires a system of work and regular review and practice to ensure skills remain sharp.

Like so many other aspects of height safety, when the accident happens the plan must already be in place. Finding a worker suspended unconscious from a lighting truss or roof beam is not the time to think about the best way of getting him down. What rescue plans exist?

The law requires a suitable and sufficient risk assessment for all work activities and to have emergency plans in hand. This should include trained personnel to carry out a rescue. You should not assume that the emergency services will be able to effect a rescue in all situations, especially within the necessary time to prevent suspension trauma and its possible fatal consequences.

Rescue of a climber suspended by fall arrest or rope access equipment or systems

Suspension of an unmoving person in a harness whatever the type can cause serious physiological problems.

These problems do not occur in the case of prolonged suspension of an active person, as a conscious person continuously changes the pressure points of the harness. In contrast, an unmoving person risks losing consciousness after only a few minutes.

The time taken for unconsciousness to occur and thus for the onset of medical problems varies according to the individual and the method of suspension used.

One thing is sure: a user who is suspended relatively comfortably in a well adjusted harness which fits correctly will be better able to face the consequences of a fall; the avoidance of trauma and pain during the arrest of a fall lessens the risk of losing consciousness when suspended in the harness.

Expert knowledge of the techniques for release and evacuation allows rescue to take place with a minimum of delay and under the most favourable conditions.

In the event of an emergency, rescuers must act quickly but safely. Progress is sometimes impeded because access is often difficult.

This difficulty requires that rescuers must have a mastery of techniques for safe and efficient passage. The selection and the installation of anchors are of paramount importance when effecting a rescue.

The variety of situations encountered means that equipment must be strong, safe, adjustable and lightweight to suit the requirements of different situations.

The downward evacuation of subjects must always be carried out safely and swiftly. These evacuations require well-trained specialised teams who can carry out the operation. For outdoor situations, the evacuation must be possible in any weather conditions.

While the downward evacuation requires the use of lifting and personal protective equipment, great care and precautionary measures must be taken to avoid further accidents.

Upward evacuation is hard work and a team effort requires absolute coordination. Techniques for upward evacuation are more complex than for downward evacuation. The use of heavy equipment such as a mechanical winch may be preferable if the access is easy, whereas a hauling system made up of pulleys and other components is necessary where access is difficult.

When the above questions have been answered satisfactorily, systems have been devised and a risk assessment carried out, the work *may* be able to commence.

Basic Access Systems

Catwalks

Stairways and permanent fixed catwalks and walkways with handrails are obviously the safest way to gain access to high level work places, but always check that handrails are safe and secure and never lean on them hard unless you are secured with a work restraint belt to an independent secure point of attachment.

Mobile Elevating Work Platforms (MEWPS)

Mobile Elevating Work Platforms or MEWPS are the next in our hierarchy of controls and are available in two main formats: the Boom Type usually referred to as a Cherry Picker and the Scissor Platform which is also know as a Flying Carpet (see ‘Lift Trucks, MEWPS and Plant’).

Tallescopes

Tallescopes, in effect another form of MEWP, are widely used as a means of access but as with all work equipment, great care is required in their use, as a number of serious and fatal accidents have occurred in recent years. The manufacturer’s instructions should always be followed when using tallescopes and access equipment in general.

Whether a tallescope is appropriate for specific tasks in the prevailing circumstances shall be assessed by the person in charge prior to use.

Maintenance and Inspection

- Do not expose tallescopes or any access equipment to corrosive substances, avoid storage outdoors and keep tallescopes clean, especially moving parts.
- Lightly oil locking collars, castors and locking pins.
- Regular inspections for damage and missing parts is essential, and all inspections should be recorded in writing.

The manufacturer can and will inspect and then quote for repair if required.

(Upright U.K. Ltd., Access House, Halesford 17, Telford, Shropshire. TF7 4PW Tel: 01952 685200)

Before use, ensure the tallescope is not damaged and that locking collars and locking pins are effective. Faulty equipment must be taken out of service, labelled as faulty and secured (by padlock and chain) to prevent use.

A written report, detailing the nature of the damage/defects, should be made to the person responsible for arranging repair by the manufacturer. The tallescope must be regularly inspected for damage or missing parts and annually returned to the manufacturer for inspection and maintenance.

Inspection checklist:

- All four wheels turn freely
- All four brakes work
- All four leg extension locks work
- Both end braces are locked into position
- Base platform is fitted on ladder side of base
- Both outriggers are fitted (one either side)
- Both outriggers' extension clips are fitted and locate
- Both outriggers' extension feet move freely
- All four ladder base upright locks locate and hold
- Both ladder extension hooks locate on rungs and lock
- Ladder extension line is in good order and runs through correct pulleys
- Ladder extension moves up and down smoothly
- Lifting basket rail is fitted and locates on latch
- Tool bag is empty of all objects and is secured to pulling line

During use:

- A Tallescope should not be used or moved without a trained and competent supervisor present and acting as person in charge.
- All those using a tallescope should be competent/trained (or undergoing training) and authorised to do so.
- Users should have read and demonstrated an awareness of the company's procedures for the safe use of tallescopes and have either provided a suitable certificate of competence or demonstrated practical competence with all aspects of tallescope use to a trainer.
- Tallescopes should not be available for use by contractors, students or those who have hired a venue/facilities unless the above requirements are met.

- Do not accumulate tools and materials on the tallescope platform or in the bag.
- If equipment is pulled up to the top of the tallescope, it should always be kept within the wheelbase of the tallescope.
- Do not stand on kickboards or rails to gain height.
- Keep tools in the tool bag.
- Do not clutter the platform or working area with equipment, keep the work area clear of all obstructions (nuts, bolts, battens and tools).
- Any personnel in the working area must be informed and made aware of the scope use.
- Persons using and in the proximity of the scope when in use must wear hard hats.
- Radio communication should be used by operators in noisy environments.
- Dismantle and re-assembly of the tallescope shall be in the presence of the person in charge. Access equipment when not in use shall be secured by padlock and chain to prevent unauthorised use.
- Lighting levels should be adequate to ensure visibility, taking into account the problems of glare.
- Obstructions on walls and ceilings, the positions of trailing wires, electrical switches and lamps should be taken into account and, in particular, precautions taken to avoid electric shock injury.
- Never go over a trapdoor without checking it is strong enough to take the load, and then observe extreme caution when doing so. Observe extreme caution spanning a gap on the stage or other working areas.
- Revolve operators and elevator operators should be made aware of 'scope use. Isolate stage machinery where possible. In particular, never have one end of the tallescope on a moving truck, revolve stage or elevator and the other on fixed flooring, without the operator's knowing there is scope activity.

For several years now there has been a dispute over the correct use of tallescopes, to be more specific, if it is safe to move a tallescope with an operator in the basket. The reason for wanting to move a tallescope with an operator in the basket is because of the fatigue factor brought on by having to climb down from the basket and then re-climb to the basket each and every time the tallescope is moved, for example; when focusing a large lighting rig.

The view of the manufacturers is that a tallescope should never be moved with an operator in the basket. Remember: following the manufacturer's instructions is a key method of controlling use. This is even more important with CE Marked equipment.

A dangerous precedent is set if we go against the manufacturer's instructions. A risk assessment will show that moving a 'scope' with an operator in the basket is a high risk that can be reduced to an acceptable level by not moving the 'scope' with an operator in the basket. The problem of fatigue can be overcome by rotating the crew; the cost of extra crew can be justified by the fact that cost can't be taken into consideration when the risks are high. This view is also adopted by the HSE.

Outriggers

Outriggers should never be removed from the tallescope, but should be folded back when in transit. Take care to loosen the clamp fitting sufficiently to prevent scoring of the vertical tubes. Repeated scoring will cause serious structural wear over time.

When in use, the width across the outriggers should be no less than one third of intended platform height. The outriggers themselves should be set at an angle of 90 degrees to the long axis of the tallescope.

Outriggers should normally be in contact with the floor at all times when the tallescope is in use but may float an inch above the ground when moving the 'scope' into position. Outrigger leg pin clips shall be properly engaged.

Do not attempt to 'tie in' the scope to a structure, to do so is a specialist decision: "Scopes should be used free standing unless under the supervision of a skilled engineer/rigger of proven competence."

Occasionally, the tallescope is used against a wall or piece of scenery, or similar. This may obstruct the use of an outrigger.

If this is the case, a member of the technical crew using the scope must be directed by the person in charge to 'foot' the tallescope on the side opposite to the wall, scenery or similar for stability. The outrigger should be reset as soon as possible and placed in contact with the floor or stage.

Use of the Tallescope

Under no circumstances shall any person be carried upon the tallescope or be moved with a person on the tallescope ladder or platform. The tallescope shall be for static use only. The tallescope should not be repositioned with materials on the platform.

Repeated climbing of the tallescope causes fatigue, so roles shall be rotated amongst the technical crew using it.

- Follow the manufacturer's instructions precisely.
- Ensure the tallescope is level. Adjustable legs must not be used to gain height and at least one, preferably two – and ideally four – of the legs must be fully retracted.
- If used on a rake, the long axis of the tallescope should be up and down the slope of the rake.
- Before climbing the tallescope, ensure that both outriggers are fixed and locked into position so that they make firm contact with the floor.
- Ensure the leg locks and ladder hooks are engaged and casters locked.
- When located on a stage, all tallescope wheels should at all times be at least one metre from the stage edge.
- Never sit on the side of the scope with someone at the top.
- On a raked surface with someone in the basket, two people should be positioned at the base, one either end. A member of the technical crew shall remain at the base of the tallescope to guard against others inadvertently moving the ladder whilst a person remains working above.
- The platform height shall be set to a level that permits the specific tasks to be completed without the user having to over-reach.
- When working upon the platform the safety bar should be closed. Kick-plates or rails should not be stood on to gain height. Never use packing, rostra or shims to gain height or level.
- The maximum platform load should be known and clearly marked. A suitable tool bag should be supplied and used. Tools and materials should not be accumulated on the platform

Equipment weighing more than 10kgs should be raised via a rope and pulley set attached to the grid and operated by the technical crew below. It should not be hauled up the tallescope.

Lifting a Scope onto a Raked Stage or Higher Platform

This should be done with the ladder in a horizontal position; this helps to keep the centre of gravity as low as possible and aids balance while lifting.

The staffing level can vary from two for a four inch or less lift, to four for a lift greater than four inches or even six or more people for an unusually high lift. Where possible, one end of the scope should be on the ground and the narrow end lifted onto the stage first.

Repeated rough handling ('beasting') *will* cause premature weakening and loosening of the structure of the scope. A planned, gentle and efficient lift will preserve the safe working life of the equipment.

If there are any reasons that prevent exact adherence to the above working systems then additional safety measures may be necessary.

Access Towers or Tower Scaffolds

Tower scaffolds come under the same level of the hierarchy as MEWPS; they can be erected quickly and can give good safe access. They are, however, involved in numerous accidents each year. These accidents usually happen because the tower has either not been erected properly or has not been used properly.

The Prefabricated Aluminium Scaffolding Manufacturers' Association (PASMA) has produced a Code of Practice for the use of tower scaffolds. Readers are advised to only purchase tower scaffolds from manufacturers who are PASMA Members and then to follow their Code of Practice and the manufacturer's instructions on its use.

Aluminium access towers should have B.S. 139 Part 3 1983 approval and be marked with the B.S. Kite Mark or a more up-to-date CE mark and standard. In all cases, the manufacturer's instructions should always be followed when erecting, using or dismantling tower scaffolds.

The instruction manual should be available on site at all times and if the scaffold is hired, the hire company ought to provide this information, and even demonstrate the safe method of building, use and striking.

If a tower scaffold is to be used:

- The tower must be vertical and legs should rest properly on firm, level ground.
- Lock any wheels and outriggers – base plates provide greater stability if the tower does have to be moved.
- Provide a safe way to get to and from the work platform, such as using internal ladders. Climbing up the outside may pull the tower over and is a definite no-no!
- Access to the working platform must be provided by purpose-built ladders or stairways which should be erected as shown in the supplier's instructions.

External ladders must never be used with aluminium towers, nor should the horizontal rungs of the tower be used as a means of access unless specifically designed for the purpose.

Certain types of tower have a vertical ladder incorporated into the end frame structure. Where these are not continuous, an intermediate platform must be positioned at the foot of the ladder.

Access to or through fully decked platform levels must be via a hatch, which must be capable of being secured in the closed position.

If independent ladders are used vertically, they must be positioned internally and fended off the frames to give adequate clearance for hands and feet. They should be firmly secured to the frames and must not rest on the ground.

Where the vertical distance between intermediate or working platforms exceeds 9 m (30ft), a rest platform with guardrail must be provided.

Platforms from which a person could fall and suffer injury must be protected by guardrails or other suitable barriers.

Barriers or guardrails must also be fitted on any intermediate platforms being used as working platforms or for storing materials.

Alloy system tower guardrails should be fitted to ensure they cannot move horizontally (usually by fitting the snap-hook to a vertical/standard, not to a horizontal rail tube) if they are leant or pulled on. Accidents can be caused by a guardrail moving during normal use.

Access towers should be rigidly tied to the structure to provide additional support if heavy materials are lifted outside the tower or if the tower is sheeted. Sheeting greatly increases the effect of wind loading on the tower, thus reducing the stability.

Be cautious about the use of towers in open-ended buildings or on outdoor stages as the wind forces in such locations can often be far greater than if the towers are used outside the building, due to the funnelling effect of the wind. Consult the manufacturer's instructions for full details of wind loading.

If any cladding, banner, drape or other material (advertising banners for example) that will act as a 'sail' are used with or fixed to a mobile alloy tower, suitable and sufficient guying must be used. A structural engineer will need to calculate this is in accordance with the appropriate code of practice.

- Do not use a ladder footed on the working platform of an access tower or apply other horizontal loads which could tilt the tower.

- Do not overload the tower or the working platform: the manufacturer's instructions will give the safe working loads a tower can support. Generally speaking, this will show the safe working load that can be supported on any platform and the safe working load that can be supported from the tower as a whole, i.e. the sum of working loads from several different platforms and safe working loads on the castors. The castors will have safe working loads stamped upon them. If required, a notice should be erected at the base of a tower showing the safe working load to all who may use the tower.
- Towers must never be moved with men or materials on the platforms. It is hazardous and illegal to move towers by pulling them along from the working platform or by the use of a powered vehicle. Never move towers in high winds and take care to avoid overhead power lines or other obstructions.

Towers should only be moved by the application of manual effort at or near the base of the tower. Ensure sure that any holes, ducts, pits or gratings are securely covered.

Care, Maintenance, Handling, Storage and Transport

The life of aluminium alloy access equipment will be increased if proper care is taken during handling, transportation and storage. Equipment should not be dropped or jarred, hammered or levered. Parts should fit with relative ease. If it is considered that a ladder, steps or access tower, etc. has been damaged, it should be withdrawn from service. Improvised repair or modification shall not be permitted. A thorough examination should be undertaken and appropriate action taken where necessary such as repair by a competent person or scrapping.

Ladders and access equipment should be capable of being individually identified. Apart from inspection before and after normal use, they should be examined regularly by a competent person. Equipment found to be defective should be suitably labelled or marked and withdrawn from service until repaired.

The inspection should include rungs, crossbars, and stiles for defect, rung to style connections, ropes, cables and all fittings, locks, wheels, pulleys, rivets, screws and hinges. The mechanism for locking hooks, adjustable legs and castors should be lubricated with a suitable lubricant.

All metal parts should be checked for twisting, distortion, oxidisation, corrosion and excessive wear, especially on treads. Broken or loose rungs, defective tie rods and broken rivets, loose hinges or other defects should be properly replaced.

Adjustable legs should be checked to see they are not bent or the treads damaged; leg adjustment securing devices should be checked to see that they operate effectively.

Frames should be checked to see the members are straight and undamaged.

Spigots should be straight and parallel with the axis of the column tube and the device for locking frames together should be checked to see that it is functioning correctly. Ancillary parts, such as outriggers and stabilisers, should be checked for damage and effective functioning of hooks and couplers.

Castors should be checked to see that each castor housing and wheel/tyre is not damaged, that the wheel rotates effectively, that the castor swivel rotates effectively and that the brake functions properly.

Platforms should be checked to see that they are undamaged and that the frames are square and true. Plywood decks should not be split or warped and should be firmly fixed to the frames.

Aluminium access equipment should not be painted or treated in such a way as to conceal defects. Any instruction signs should be checked and replaced as necessary. A record should be kept of these inspections.

Storage areas should be easily accessible and protected from the elements. Ladders should be stored on racks designed for their protection when not in use. The racks should have sufficient supporting points to prevent sagging. Materials should not be placed on stored equipment. Ladders should not be hung from the style or rung.

During transport, ladders should be properly supported to avoid sagging and there should be minimal overhang between points. They should be tied to each support point to minimise rubbing and the effects of road shock.

Other equipment should be carefully loaded so ladders are not subject to shock or abrasion. Space can be saved by systematically placing braces, platform stairways, etc. in available space within vertically stacked frames.

Forklift Trucks with Work Platforms

Next down the list in our hierarchy come forklift trucks fitted with a work platform. Forklift trucks must never be used to lift people unless fitted with the correct type of work platform.

A MEWP should always be used in preference to a forklift fitted with a work platform (see chapter on Forklifts, MEWPS and Plant).

Ladders, Zarges, Step-ladders and Trestles

Each year around 20 people are killed at work while using ladders, and more than 1500 are seriously injured. Ladders and steps are so commonplace in most work situations that people often underestimate the risk involved with their use. Simple precautions would prevent the vast majority of ladder accidents and the resulting deaths, injuries, and suffering, not to mention financial losses.

Before selecting a ladder as a means of access, a risk assessment should be carried to demonstrate that a more suitable means of access is not justified because of the low risk, the short duration of the work and existing features on site which cannot be altered (**Schedule 5** Work at Height Regulations – Requirements for ladders).

For our purposes these items of equipment should never be made of wood – aluminium is probably the safest material. It's a little known fact that there are two standards available: a domestic standard and a more heavy-duty industrial standard. Your local DIY store will almost certainly stock the lightweight domestic standard; avoid these at all costs, they are not up to the hammering they are going to get 'on the road'. Zarges and step-ladders must be CE marked and meet the standard set by EN 131.

Ladders

Ladders are best used as a means of getting *to* a workplace; they should only be used *as* a workplace for short-term work. They are only suitable for light work and are covered under **Schedule 5** (Requirements of Ladders) of the Work at Height Regulations.

Make certain there is no better means of access before using a ladder, for this type of work is dangerous and many accidents take place during work lasting 30 minutes or less.

Ladders should be thoroughly inspected by a 'competent person' on at least an annual basis and a record kept of the inspection in accordance with **Regulation 12** of the Work at Height Regulations. Items known as 'ladder tags' are available for marking ladders with details of inspections, etc. These items are permanently attached to the ladder and are to be recommended. The user should also inspect the ladder before use.

The foot of any ladder should be supported on a firm and level surface and should not rest either on loose material or other equipment to gain extra height.

Attachments for levelling up feet on sloping surfaces should be properly fixed and used.

In no case should the bottom rung be placed so that the total weight is carried on the rung; only stiles are designed for this purpose. Only one person at a time should be climbing the ladder.

The head of the ladder should rest against a solid surface able to withstand the imposed loads. Where the surface may be fragile or brittle so that it cannot withstand such loads, equipment such as 'ladder stays' must be supplied and used.

It must be ensured that a ladder cannot slip and, wherever practicable, the top should be securely fixed.

Slips may be prevented by the use of a lashing, strap or proprietary clip secured to both stiles or, where suitable, by equipment such as restraining straps or tensioned guys. On slippery floor surfaces special care is necessary to prevent the ladder foot from moving.

Where securing at the top is impracticable, arrangements must be made to prevent the ladder from slipping outwards or sideways.

Methods of securing at the base include fixed blocks or cleats, sandbags or stakes embedded in the ground. Additionally, to help prevent slipping, most ladders can be fixed at the foot with pads, caps or sleeves.

While lashings, etc. are being fixed, or in circumstances where it is impracticable to fix the ladder at the top or foot, a second person should be stationed at the foot to prevent it from slipping. This precaution, however, is considered to be effective only for ladders not more than 5m (16ft) in overall length. The person 'footing' should face the ladder with a hand on each stile and one foot resting on the bottom rung.

The stepping-off rung of a ladder should be level with the platform. Ladders should be extended to a height of at least 1.05m (3ft 6inches) above the landing place or above the highest rung on which the user has to stand, unless there is a support handhold to provide equivalent support. This is necessary to reduce the risk of overbalancing when stepping off at the top.

The ladder should be placed at a suitable angle, ideally about 75 degrees to the horizontal, i.e. about 1 m out of every 4 m in height or a 1:4 ratio. The user should face the ladder when climbing or descending.

Where a ladder rises a vertical distance of 9 metres or more above its base, suitable safe landing areas or rest platforms must be provided at regular intervals.

A ladder (like any piece of Work Equipment) should only be used for the load and purpose for which it is designed. For example, a ladder should not have scaffold boards laid on its rungs and should not be used as an upright of a ladder scaffold unless it is heavy duty and capable of carrying the loads imposed.

The rung of an ordinary ladder is designed to support the weight of a person and whatever light tools they may be carrying, but not the additional weight of a ladder scaffold.

Ladders must not be used in areas where live conductors are exposed and there is a risk of shock or short circuit. Fault-finding, checking for blown fuses, etc. should not come into this category if live conductors are shrouded where possible and a competent person is assigned to the task.

It is important that mud and grease, etc. is cleaned off footwear before any attempt is made to climb a ladder. When ladders become contaminated, they should be taken out of service and cleaned. There should be sufficient space between the rung to provide a proper footing.

It can be dangerous for a person to carry loose tools manually up or down a ladder because he/she may be unable to grip the rungs; this is one of the most common causes of overbalancing. Light tools should be carried in a holster attached to a belt, or in a tool bag.

Other tools and materials should be raised or lowered on a rope. Ladders must be used in such a way that a secure handhold is always available to the user and the user can maintain a safe handhold even when carrying a load.

Ladders should always be provided to give a safe means of access and egress to lighting, sound and follow spot towers so that operators are not required to climb scaffolding.

Sections of extension ladders should overlap by a minimum of:

The user should raise and lower the ladder from the base and should ensure that the hooks are properly engaged. The rung 1.05m (3 ft 6in) from the top of a single section or an extension ladder is the highest to be used for climbing.

up to 5m (16ft) closed length	1.5 rungs
between 5m (16ft) and 6m (20ft) closed length	2.5 rungs
over 6m (20ft) closed length	3.5 rungs

No interlocking or extension ladder shall be used unless its sections are prevented from moving relative to each other while in use. Mobile ladders must be prevented from moving before they are stepped on.

The height for which a ladder will be unsuitable for use depends on the space available, the nature of the work, the physical effort required to erect the ladder and the cost involved, such as if more than one person is needed to erect it.

While two people may be able to handle a ladder longer than 10m, the weight involved may cause strain injury.

When not in use, ladders (and all access equipment) must be secured to prevent unauthorised use. Ladders should be stored flat and if possible raised on blocks at each end with extra support in the centre.

Zarges, Step-ladders and Trestles

Step-ladders and trestles are not designed for any degree of side loading and this should be avoided. They should be spread to their fullest extent and properly levelled for stability and should be placed at right angles to the work whenever possible, on a level surface. Work should never be carried out from the top platform, nor should overhead work entail overreaching.

The top tread of a pair of steps, bucket or tool shelf should not be used for foot support unless there is an extension above the top to provide a handhold; rear parts of steps should not be used for foot support.

Step-ladders are prevented from spreading by means of stays, chains or cords. These should be of sufficient and equal length, kept in good order and should be renewed if found to be defective.

Only one person should use a step-ladder at any one time and if steps are used in a doorway, the door should always be wedged open securely.

Trestles are made with a swing-back similar to step-ladders, but both halves have heavy cross-bearers to support a working platform. Platforms should be of lightweight staging.

Access to trestle platforms should be by means of a step-ladder. Access to fixed or unsupervised ladders should be prevented by lockable anti-climb guards or by preventing access by other means.

Truss Access and Fall Arrest Systems

Accessing and working on a truss is not a recommended system of work. Before using truss ladders and 'walking the truss', other safer systems such as the use of a MEWP or Tallescope must be considered. Only if it is not practicable can the use of truss ladders and truss walking be considered as an option.

Good fall protection starts with fall prevention!

Anyone climbing a truss ladder is required to use fall arrest protection. This is easily achieved by the use of a piece of fall arrest equipment known as retractable (inertia) fall arrester, which consists of a drum of cable fitted with a self-locking function and an automatic tensioning and return facility for the cable. The cable from the retractable fall arrester is attached to the full body harness of the climber. (Other suitable systems such a 'self life-lining' also exist).

The retractable (inertia) fall arrester must be positioned directly above the climber and attached to a suitable point that is strong enough to withhold the weight of the climber after shock loading. It is unlikely that the truss is strong enough to take the load of a falling climber and therefore should not be used as an attachment point for a retractable (inertia) fall arrester.

Inertia fall arresters will normally be marked with the Safe Working Load of the point the fall arrester should be attached to.

Climbing above the fall arrest device is dangerous and can create forces that the fall arrest device and the supporting structure were not designed for.

It is recommended that a retractable fall arrester that locks off rather than slowly lowering be used. This prevents abuse of the fall arrester, as it has been known for operators to take an easy trip down to the deck by jumping from the truss or out of the followspot seat and allowing the fall arrester to lower them. This practice must be discouraged as it burns out the clutch type mechanism of the fall arrester. A retractable fall arrester should be attached to a point that is capable of withstanding the dynamic loads imposed upon it. Again the question is: will the truss, the rigging points or the rigging equipment withstand such a loading? (Such values are usually in the 15 - 20 kN range in manufacturer's information for safe use – that is between 1.5 and 2.0 tonnes).

The courts could find you negligent if a fall arrest system is attached to a weak point. Do you know what the possible loading is likely to be in the event of a fall?

While on the subject of truss ladders, I should mention some of the frightening ways I have seen these rigged and used. Anyone who has every tried to climb such a ladder with a hook lacing system on their boots will agree that these should be avoided at all cost. The hooks get caught in the side wires and you can end up hanging upside-down by your boots – not a recommended position!

When rigging the ladder:

a) Never hang the ladder from a rung, this will cause the rung to bend and break.
b) Never hang the ladder by joining the 'C' Links fitted at each end, use a wire belay or a device known as a spreader (or spreader belay) or other similar systems as they will distribute the load correctly. Some manufacturers produce ladder attachments that clip onto the truss and the use of these is to be encouraged.
c) Never stand below a ladder holding it tight for the climber; with the right climbing technique it is unnecessary for anyone to hold the bottom of the ladder for you.
d) Always obtain the required training and experience under controlled conditions before using any rigging, climbing, work positioning, rope access or fall arrest equipment in a work situation.
e) Wire rope ladders should perhaps comply with the American ESTA standard, since at present there is no standard set in the UK or Europe.

One well-known manufacturer of truss and rigging equipment has developed a very strong truss that incorporates a fall arrest system.

This is achieved by a patented chord extrusion which has a continuous T-slot along the top and bottom of the truss. This means that the climber may remain constantly attached to a properly designed anchor point while *moving* along the truss. The anchor points and lanyard all conform to EN 795-B, are CE marked and comply with the PPE Directive.

My interpretation of the PPE regulations (that state that the use of PPE must be suitable and sufficient) is that a worker is required to remain *continuously* attached to a suitable anchor point whilst at height, in order to be using a safe working system.

Great care must be exercised when using catenary wire/webbing systems (as running anchor or attachment points) along the tops of single trusses in particular. The system *must* be designed so that the reactions at the anchorages in the event of a fall occurring are within the design parameters of the truss being used. If the system is rigged to the chain motors or other suspension equipment, the same principle must be followed.

Using a proprietary horizontal lifeline/fall arrest system such as this on a truss or truss system requires analysis by a structural engineer as well as the approval of the truss manufacturer as 'suitable and sufficient' as PPE. The majority of truss manufacturers, when asked, will not authorise use of their trusses for such purposes.

It is not sufficient to clip a lifeline/fall arrest system onto a tensioned horizontal wire rope fixed directly to the truss or a master-link below the hoists.

It is not uncommon to see wire ropes of diameters as small as 6mm being used with no shock absorber, by riggers using sit-harnesses purchased from mountaineering suppliers.

Section 37 of the HASAWA makes interesting reading when you consider that this fall arrest information is now becoming common knowledge…

The forces generated by such equipment in a fall are typically in excess of one tonne and are clearly outside of the strength of most trusses.

A standard truss or grid system is unlikely to have been designed to resist fall arrest forces. Manufacturers of master-links or shackles are unlikely to recommend angular dynamic loading of a typical master-link already in vertical tension.

The hoist brakes or overload mechanisms may not prevent the hoist slipping, particularly when the truss is already loaded with lighting equipment, cable, and so on.

Hoist manufacturers generally require loads to be secured by appropriate secondary means if there is human access to the load.

The ground clearance required must also be understood. Many of these systems rely on the use of 'unzipping' shock absorbers and dynamic materials to reduce the shock loads on a person falling from the structure they are attached to. Ground clearance of 9m is not an uncommon requirement of manufacturers.

In many entertainment situations, a falling worker would hit the ground before the fall was arrested by the equipment.

The truss *manufacture* of the 'fall arrest truss' mentioned has achieved a minimum clearance of around 3m.

One factor that is often overlooked is the compatibility of a fall arrest system and the height at which a person is to work.

Energy absorbers are a crucial part in a fall arrest system, but their very nature can increase the ultimate length of a lanyard by an extra 1.75m (EN 355). Therefore, the total clearance between the anchor point and the ground must be taken into consideration before selecting fall arrest equipment. See table above.

Lanyard Length	Fall Distance	Safe Fall Clearance based on testing (see notes below)
1.00m	2.0m	4.6m
1.50m	3.0m	4.9m
1.75m	3.5m	5.8m
2.00m	4.0m	6.2m

This is an example of safe fall clearances required for differing length shock absorbing lanyards based on a 100kg weight. Each calculation is based on:

- Average height of person 2 metres.
- Full length of lanyard inclusive of connectors.
- Safety distance = minimum of 1m, which allows 0.5m for incorrect harness fit/take-up of slack and attachment point extension.
- Length of energy absorber after deployment derived from testing.

It must be noted that some variation may occur between different manufacturers.

All testing was carried out using a Fall Factor of 2, which is equal to the anchor point being at or near foot level and which is considered with being the worst-case scenario, but all too close to reality.

The maximum permissible extension of a shock absorber that is tested to comply with EN 355 must not exceed 1.75m. So, if a 1.75m shock absorber is to be correctly used without the possibility of serious injury then a minimum safe fall distance of 5.8m is required.

If a person is required to work at a height of 4m and the only safe anchor point available is at or near foot level, using the above table it can be determined that a standard lanyard of between 1m and 2m is inappropriate and is not suitable or sufficient to prevent serious injury.

In this situation the only option is to use a system that will allow the potential fall distance to be reduced.

Such systems include rope grabs and fall arrest blocks that must be attached to a high anchor point situated above head height whilst allowing the person to attach to the system at the harness attachment point level. This will reduce the safe fall distance in this instance to about 3.0m.

Even if the supporting structure and the ground clearance are adequate, there must still be a rescue system and trained rescuers available to immediately effect rescue.

In the ESTA *Protocol* magazine, Peter Hind, a structural engineer working in the entertainment industry, is quoted as follows:

"In this situation, the typical values for a 100 kg weight falling two metres from a line 18 metres (60ft) long would be between 1 and 2 tonnes, depending on whether shock absorbers are fitted. The question now posed is: will the truss, the rigging points (i.e. the building or stage roof) or the rigging equipment (such as motors) withstand these forces?"

Unfortunately, there are no simple answers, but the answers will need to be found before you can use such fall arrest systems safely.

The new truss system I mentioned earlier that is fitted with attachment points that slide along the length of the truss that the climber may attach to via a harness, shock absorber and connectors, goes a long way towards solving our problems, as this truss is strong enough to take the shock load of a falling climber. However, checks must be made to ensure the rigging points, motors, round slings and all other associated rigging equipment is also strong enough to withstand the weight of a dynamic fall.

Promoters and production managers need to take a very close look at risk management systems, risk assessments and the regulations to ensure that there is sufficient compliance. At present our systems are severely lacking.

Personal Protective Equipment (PPE) for Work at Height

If there is any risk of falling and all other means of safeguarding the worker such as guardrails are not possible, fall arrest equipment must be worn or safety nets, air bags or other 'collective' systems used. Safety of personnel must be of prime importance.

Standards exist for fall arrest systems, and these must be used in preference to rock climbing techniques and equipment purchased from climbing shops.

It should be understood that although a climbing harness may look as if it will do the job, it is designed for a voluntary activity.

Any work equipment must be provided by the employer (if self-employed the HSAWA requires the self-employed to carry on their business as though they were an employer) and must therefore be of suitable quality and design. The requirements of an employer differ fundamentally from those of someone going rock climbing.

The average unconscious human suspended in any type of harness has been found to risk severe and possibly fatal restriction in blood circulation (thus oxygen supply to the brain) in about fifteen minutes. This condition is known as suspension trauma.

Whilst the fallen worker is conscious, a well fitting EN361 harness can be relied on for some time before serious risks to safety arise, as the suspended person is able to move a little to avoid the pressure of the harness.

If a 'safety' system is employed, it must be safe. There is only one time that the security of the system will be tested.

Many systems require design by engineers and with input from manufacturers and suppliers. They must be inspected and treated like any other PPE.

All PPE for work at height is Type 3 (Complex), that is to say it protects from mortal danger.

Type 3 equipment must:

1) Undergo independent type testing usually to European Standards (ENs).
2) Have appropriate technical and user instructions.
3) Be produced under an independently verified quality system or be subject to periodic batch testing.

At the time of writing a new British Standard is currently being drafted for the Selection, Use and Maintenance of Fall Protection Systems.

Helmets

A risk assessment will identify the need for any worker at height using fall arrest equipment to wear head protection to guard against being knocked unconscious in the event of a fall.

This will reduce the risk of death in a similar way to the use of the correct harness standard. The head protection will take the form of a lightweight helmet that protects the head from impact from objects the user may fall against. This will be a side and a vertical impact. The helmet will therefore require a restraining strap cradle that is designed to account for these forces and to prevent the helmet being displaced in a fall.

This will usually occur when the brim strikes an edge as the wearer falls or brushes past an object. The cradle must fail before there is a risk of strangulation if the user was entangled and suspended by the helmet.

The only type of helmet which currently meets our requirements is the one used by rock climbers: it should be worn by those working at height.

The helmet should carry a CE Mark and comply with EN 12492: 1996, but

I must mention that this is a standard for leisure and recreation, no EN standard exists for the industrial use of climbing helmets. However, climbing helmets such as these are already in widespread industrial use and can be safely adopted in our industry.

This type of helmet is preferable as it is designed to absorb the impact to the head when the wearer falls from height. It is also fitted with a chin strap and head cradle that will keep the helmet on the user's head in the event of a fall. Standard industrial safety helmets (EN 397: 1995) must be worn by those who, out of necessity, need to remain below a person working overhead.

A well-known French manufacturer of caving, climbing and safety equipment (whose products are readily available in the UK) now produces a helmet that conforms to both EN 12492:1996 and EN 397:1995. This helmet will become the standard helmet for our industry if it has not done so already – but there will be others…

The regulations do not require climbers and riggers to wear steel toe capped footwear when climbing but of course, when involved in 'non-climbing' work and more precisely manual handling they are subject to the same regulations as everyone else and must use protective footwear.

Do I Need a Harness? If So, What Type?

This will depend on the type of work you are doing. Whatever type of work the harnesses should be chosen with great care and the worker should also be involved in the selection. For instance, if the harness is to be used with the operator suspended for long periods, then a comfortable well-padded full body harness will be required, thus involving the worker in careful selection is vital. There are three situations where a harness or belt is required:

a) Fall Arrest:

A *full body* harness is required for anyone working *two metres* or more above the ground where there is a possibility of the worker falling. In addition to a harness, the provision of a 'shock-absorbing sling and hook' and the requirement to be "attached at all times" to a suitable anchor point must be complied with.

The requirement for Fall Arrest is to use a system that complies with EN 363 consisting of:

1) A full body harness to EN 361: 1993
2) An energy absorber to EN 355: 1993
3) A lanyard (i.e. slings) to EN 354: 1993

4) Connectors in the system (i.e. karabiners and hooks) to EN 362: 1993
5) An anchor capable of sustaining the dynamic load in the event of a fall, also to EN standards

The standards used for fall arrest all aim to reduce the impact force on the body and the anchorages of the system to below 6 kN. This requires the *complete* safety system to conform to the standards set out and to be used correctly.

The shock-absorbing lanyard used for fall arrest must be connected to the upper chest or dorsal attachment point on the harness. The side ('pole strap') and waist dee rings are not for fall arrest.

Simple work positioning (reducing risk further by limiting exposure to situations where a fall could occur in the first place) can be used in addition. To use fall arrest lanyards with a work positioning harness ('sit' or 'abseil' harness) is dangerous.

The maximum length of a lanyard (including shock absorber and connectors) is two metres. The shorter the lanyard the better and always attach yourself to a point above you not below as this will increase the fall factor. Twin-tailed lanyards are preferred as they allow the worker to remain attached whilst moving from one position to the next.

Two types of full body harness are available: a simple system for basic fall arrest and more complex multi-purpose systems that can also be used for work positioning and work restraint. Some are fitted with dee ring to attach a work positioning pole strap together with padding on the waist and leg straps to make long duration work positioning more comfortable.

b) Work Positioning:

A sit harness (to EN 813: 1997) may be used for work positioning (i.e. rope access work such as abseiling, or prussicking into position for work) but not for fall arrest.

Much of the work carried out by 'riggers' in our industry involves mixed working situations whereby one minute a rigger may be climbing out on a beam or truss (fall arrest) and the next minute abseiling from the roof from that beam or truss (rope access). So what sort a harness does he need? Well, the answer is a harness that fits both requirements and the highest standard required; for fall arrest it must be a full body harness.

c) Work Restraint:

A waist-belt or climber's harness may be used for restraint purposes, that is to say, to *prevent* a person reaching zones where the risk of a fall exists (e.g. a high platform or stage with no safety rails). It is accepted practice that protection is used when working within two metres of an edge.

No shock absorber is required in a Work Restraint system, the belt or harness must be fastened according to the manufacturer's instructions, and a sling and karabiners used to attach the user (via the belt) to a secure anchorage. The system must be used correctly and 'so far as is reasonably practicable' does not permit and cannot be adjusted to permit incorrect use. [**Schedule 4**, Part 5].

Work restraint equipment must comply with the following regulations: EN 358: 1993 for Waist Belts (for Work Restraint only), EN 354:1994 for Lanyards and nylon webbing slings and EN 362: 1993 for any connector in the system (e.g. a karabiner or hook).

So, when climbing, such as during rigging operations, a full body harness for 'fall-arrest and work positioning' purposes is now the requirement of the enforcement agencies. Hence this type of product should be purchased for rigging rather than waist belts or sit harnesses, etc.

Following an accident involving a climber involved in rescue training, the HSE issued a warning to climbers and riggers about the use of karabiners fitted with a three-way action locking mechanism. They may be unsafe to use as a main method of attachment as they may become detached from the rope system; my view is that this can happen with any karabiner. One danger is with the use of so-called 'self-locking' karabiners: people assume that because it is self-locking, they don't need to check to see if the gate is properly secure. My advice is to always check the gate is securely fastened, regardless of the type of karabiner.

Anchor Points

Once we have our FPE (Fall Protection Equipment) that is most suited for the intended application, we must consider suitable anchor points to which we can attach the equipment. Anchor points form the basis of the whole system, so poor selection will result in total failure of any fall arrest system.

Anchor points come in several types, they may be natural, manmade, permanent or temporary.

Anchor types:

1. *In situ*

2. Class A1 and A2
3. Class B
4. Class C
5. Class D
6. Class E

1. In situ

An anchorage, either natural or manmade, that is of a permanent nature, i.e. rock boulders or structural brickwork/steel, to which an anchor device or personal protective equipment can be attached.

Selection of an *in situ* anchor is viewed as personal preference, as there are no standards or guidance on the subject from a legislative view. The user must consider the following points when selecting an *in situ* anchor:

a) Ensure the anchor is of suitable strength and integrity to be able to take a shock load of a minimum of 1000kg. If the anchor point is not strong enough it may be possible to share the load between two or more points.
b) Ensure the anchor is compatible with the system and that the system can attach to the anchor.

2. Class A1 and Class A2

Class A1 is a structural anchor designed to be attached to vertical, horizontal and inclined surfaces, e.g. eye bolts.

Class A2 is a structural anchor designed to be attached to an inclined surface, i.e. anchor plates.

The standard for Class A1 and A2 is: Protection against falls from a height – Anchor devices – Requirements for testing EN 795: 1997

Code of Practice for the Application and use of anchor devices conforming to EN 795 – BS 7883: 1997.

Class A1 and A2 anchors are made in a variety of styles, shapes, sizes and materials to meet different use demands. It is important to ensure that any selected anchor is compatible with the Fall Protection Equipment system. Points to consider are:

a) Does the anchor conform to EN 795: 1997? This will be marked on the anchor or on the packaging.
b) Is the anchor compatible? Can you connect the system to the anchor?
c) In what direction will the impact load be imposed?

Due to the number of different Class A1 anchors it is important that the correct one is used. For example, some eye-bolts can only take a shock load in line with their longitudinal axis.

d) Is the constructional material adequate? The material in which the anchor is attached must be sufficiently stable and provide adequate security against collapse if an arrest force is applied. The installer should undertake an assessment of this.
e) Has the anchor been installed correctly? For expanded socket or chemical-bonded anchor, the installer must apply an axial force 5kN for 15 seconds to confirm soundness.
f) Is there a test certificate? The person who installs the anchor should provide a certificate to show the anchor has been installed in accordance with the code of practice for application of anchor devices conforming to EN 795, BS 7883: 1997. The certificate must include a warning against misuse of the anchor and draw the attention of the user to the need to inspect the anchor before each use.
g) Inspection and examination. Each anchor device should be visually inspected and manually checked before each use.

At least every 12 months a 'competent person', authorised by the manufacturer, should examine each anchor device in accordance with the manufacturer's instructions. Class 1A anchors are re-certification after the 12-month inspection has been carried out; re-testing may be required before this can take place.

3. Class B

Class B is a temporary transportable anchor such as a beam trolley or girder clamp.

The standard for Class B is: Protection against falls from a height – Anchor devices – Requirements for testing EN 795: 1997

Code of Practice for the Application and use of anchor devices conforming to EN 795 – BS 7883: 1997.

Class B anchors are made in a variety of styles, shapes, sizes and materials to meet different use demands. It is important to ensure that any selected anchor is compatible with the Fall Protection Equipment system. Points to consider are:

a) Does the anchor conform to EN 795: 1997? This will be marked on the anchor or on the packaging.

b) Is the anchor compatible? Can you connect the system to the anchor?
c) Has the anchor been positioned correctly? Ensure the anchor is in a stable position for use. Consideration must be given to the structure to which it is attached or positioned upon, to ensure that it is capable of taking any load it may be subjected to. This may require consultation with a structural engineer. If the anchor is to be used in a number of different positions, ensure each position meets the requirements and that each user is aware of the acceptable positions.
d) Inspection and examination. Each anchor device should be visually inspected and manually checked before each use. At least every 12 months, a 'competent person', authorised by the manufacturer, should examine each anchor device in accordance with the manufacturer's instructions.

4. Class C

A Class C anchor device employs a horizontal flexible line, for example a horizontal safety line.

The standard for Class C is: Protection against falls from a height – Anchor devices – Requirements for testing EN 795: 1997

Code of Practice for the Application and use of anchor devices conforming to EN 795 – BS 7883: 1997.

Class B anchors are made in a variety of styles, shapes, sizes and materials to meet different use demands. It is important to ensure that any selected anchor is compatible with the Fall Protection Equipment system. Points to consider are:

a) Does the anchor conform to EN 795: 1997? This will be marked on the anchor or on thc packaging.
b) Is the anchor compatible? Can you connect the system to the anchor?
c) Is the material compatible with the environment in which it is to be used?
The material of the horizontal flexible line is not going to be affected by the environment in which it is to be used. Nylon is affected by acids and polyester is affected by alkalines.
d) Is the constructional material adequate? The material in which the anchor is attached must be sufficiently stable and provide adequate security against collapse if an arrest force is applied. The installer should undertake an assessment of this.

e) Has the anchor been correctly installed? The installer of the anchor must ensure the anchor is tested in line with EN 795.
   Also, in calculating deflection that will occur in the system under load, that any user having fallen will not make contact with the surfaces or other objects.
f) Is there a test certificate? The person who installs the anchor should provide a certificate to show the anchor has been installed in accordance with the code of practice for application of anchor devices conforming to EN 795, BS 7883: 1997.
   The certificate must include a warning against misuse of the anchor and draw the attention of the user to the need to inspect the anchor before each use.
g) Inspection and examination. Each anchor device should be visually inspected and manually checked before each use.

At least every 12 months a 'competent person' authorised by the manufacturers should examine each anchor device in accordance with the manufacturer's instructions.

5. Class D

A Class D anchor employs horizontal ridged anchor rails, for example horizontal Suretrack.

The standard for Class D is: Protection against falls from a height – Anchor devices – Requirements for testing EN 795: 1997

Code of Practice for the Application and use of anchor devices conforming to EN 795 – BS 7883: 1997.

Class D anchors are made in a variety of styles, shapes, sizes and materials to meet different use demands. It is important to ensure that any selected anchor is compatible with the Fall Protection Equipment system. Points to consider are:

a) Does the anchor conform to EN 795: 1997? This will be marked on the anchor or on the packaging.
b) Is the anchor compatible? Can you connect the system to the anchor?
c) Is the constructional material adequate? The material in which the anchor is attached must be sufficiently stable and provide adequate security against collapse if an arrest force is applied. The installer should undertake an assessment of this.
d) Is there a test certificate? The person who installs the anchor should

provide a certificate to show the anchor has been installed in accordance with the code of practice for application of anchor devices conforming to EN 795, BS 7883: 1997. The certificate must include a warning against misuse of the anchor and draw the attention of the user to the need to inspect the anchor before each use.

e) Inspection and examination. Each anchor device should be visually inspected and manually checked before each use. At least every, 12 months a 'competent person', authorised by the manufacturer, should examine each anchor device in accordance with the manufacturer's instructions.

6. Class E

A Class E anchor is a dead weight on a horizontal surface, for example a weight trolley.

The standard for Class E is: Protection against falls from a height – Anchor devices – Requirements for testing EN 795: 1997

Code of Practice for the Application and use of anchor devices conforming to EN 795 – BS 7883: 1997.

Class E anchors are made in a variety of styles, shapes, sizes and materials to meet different use demands. It is important to ensure that any selected anchor is compatible with the Fall Protection Equipment system. Points to consider are:

a) Does the anchor conform to EN 795: 1997? This will be marked on the anchor or on the packaging.
b) Is the anchor compatible? Can you connect the system to the anchor?
c) Is the surface adequate? The surface on which the anchor is placed must be strong enough to prevent collapse. A competent person who may be a structural engineer must take an assessment of this. It is important to follow the manufacturer's advice that should be marked on the product.
d) Has the anchor been installed correctly?
The installer of the anchor must ensure that it is positioned and weighted so as to provide the security that is required for the safety of all users. A competent person who may be a structural engineer should determine this.
f) Inspection and examination. Each anchor device should be visually inspected and manually checked before each use. At least every 12

months a 'competent person', authorised by the manufacturer, should examine each anchor device in accordance with the manufacturer's instructions.

Care of PPE

PPE, including slings, karabiners, connectors, harnesses and belts must never be used for lifting or hauling purposes (apart from in rope access systems). This would not only be dangerous, but is also inappropriate use of the equipment under both the Provision and Use of Work Equipment Regulations and the PPE Regulations.

Equipment used for rigging, such as Span Sets and shackles, should be kept separate from fall arrest, work positioning and work restraint equipment.

Climbing helmets must be treated and looked after like any other safety helmet. They must be kept out of strong sunlight, as the ultra violet rays can destroy the polyamides the helmet is made of without the user knowing. If the helmet is dropped from height or receives a severe blow it must also be scrapped, since hairline cracks invisible to the naked eye may be the result of such an impact and can weaken the helmet to a critical degree.

Work Restraint belts, harnesses and slings need special care. Like all equipment for work at height, lives depend on these items.

Never buy second-hand PPE and do not share any items of PPE.

Employers have a duty to issue free of charge new PPE to staff who require such equipment.

Employees have a duty to use PPE correctly and to take care of any PPE issued to them. Loss, damage and faults must be reported and new equipment then issued. A collective 'pool' of helmets and harnesses, etc. is not acceptable (see Personal Protective Equipment Regulations).

Karabiners should always be visually inspected by the user before use. If they are suffering from wear, defects, cracks, abrasion, burrs, distortion/deformation, corrosion or contamination by chemicals, or if they have been subjected to a fall or dropped from height, they should be scrapped.

After a fall or drop, hairline fractures can develop (that are only detectable with an electron microscope) that can weaken the equipment to a dangerous extent. If in doubt, scrap it and get a new one! To prevent others from using scrapped and unsafe equipment, it should be cut or sawn up before it is disposed of.

These items can be cleaned in warm water and the threads of karabiners can be given a very light lubrication with silicone grease or light oil from time to time. Karabiners should be stored unpacked in a cool, dry, dark place away from heat sources, high humidity, sharp edges, corrosives, or other causes of potential damage.

A screw gate karabiner should always be used with the gate screwed up. The loading and test figures for karabiners are calculated with the gate screwed up. A karabiner loses 50% of its strength when the gate is undone.

Never join two or more karabiners, because they can twist and unclip themselves.

Never apply a 'three way loading' to a karabiner and never load a karabiner across its minor axis. They are designed to be loaded one way only – along the major axis.

After use, harnesses, work restraint belts, lanyards and slings should be coiled and hung up in a cool, dry, dark place from a wooden peg or from a loop of string or cord. They should never be hung from metal pegs or nails, nor should they be stored near high heat sources, corrosives, chemicals, oils, solvents, sharp edges, high humidity, direct sunlight and other causes of potential damage. UV light (from sunlight or UV lamps) can considerably weaken harnesses, belts, lanyards, slings and span sets. Take care where UV lighting is used on stage and cover span sets to protect them from the unseen effects of UV lighting.

Destroy any harness, belt, lanyard or sling that comes into contact with chemicals or if contact is suspected. As a general rule, any chemical within the pH range of 5.5 to 8.5 will probably be safe, as will oil-based chemicals, but always check!

Harnesses, belts, lanyards and slings that become dirty should be washed in warm water (max 40° C) with pure soap or a mild detergent (within a pH range of 5.5 - 8.5). UK examples are Lux flakes or Stergene. The use of a washing machine is permissible, but place the equipment in a suitable bag to protect against mechanical damage.

A synthetic scrubbing brush may be used on heavily soiled harnesses, belts, lanyards or slings. After washing rinse in clean, cold water. The harness, belt, lanyard or sling must then be hung up to dry, as previously described.

Any fall arrest or work restraint equipment involved in a serious fall must be scraped after the accident investigation has been completed. A competent person should thoroughly examine any equipment that has been subject to a minor fall and discard it if there is any sign of defect or any doubt about its safety.

As a basic guide, in normal use, slings and harnesses have a storage life of 10 years and a maximum working life expectancy of five years. Any harness or lanyard that has been submitted to a fall must be scrapped.

The PPE and the Work at Height regulations require inspections on all PPE and work at height equipment to be made by a *competent person* (who has experience and, in some cases, has received the required training and authorisation by the manufacturer).

All equipment must be inspected by the user every time the equipment is used; this will normally mean a simple daily inspection regime that (with the exception of MEWPS) does not need to be recorded. During inspection particular attention should be paid to cuts, tears and abrasions, damage due to deterioration, contact with heat, acids or other corrosives and chemicals. Check sewing for broken, cut or worn threads.

The purpose of an inspection is to identify whether the equipment is fit for purpose and can be used safely, and that any deterioration is detected and remedied before it results in unacceptable risks. An inspection can vary from a simple visual or tactile check to a detailed comprehensive inspection, which may include some dismantling/testing. A competent person should determine the nature, frequency and extent of any inspection, taking account such factors as the type of equipment, how and where it is used, its likelihood to deteriorate, etc. For example, if equipment is to be used in onerous outdoor conditions, it may need more regular inspections than similar equipment used indoors. Periods between inspections should be chosen on the basis of risk assessment, and should be reviewed in the light of experience.

Equipment covered by the Lifting Operations and Lifting Equipment Regulations (LOLER) (such as rope access equipment used for abseiling, etc.) will also be subject to a thorough examination regime that does not have to be repeated under the WAHR.

Regulation 12 (4) requires that a weekly inspection is carried out for scaffolding, as previously required by the Construction (Health, Safety and Welfare) Regulations (CHSWR).

Where work equipment is hired to the user, it is important that both parties agree, in writing, exactly what inspection has been carried out and that information is available and can be passed to the workers.

Inspection and thorough examination are not a substitute for properly maintaining equipment. The information gained in the maintenance process, inspection and more technical thorough examinations, should be shared and the processes should be complementary. If a maintenance log exists, make sure it is kept up to date and accessible to the competent person performing the inspection or thorough examination. The maintenance process also needs proper management:

> Planned preventative maintenance involves replacing parts or making necessary adjustments at pre-set intervals so that risks do not occur as a result of the deterioration or failure of the equipment; and
>
> Condition-based maintenance involves monitoring the condition of safety-critical parts and carrying out maintenance whenever necessary to avoid hazards which could otherwise occur. This would include, for example, hydraulic systems in a MEWP.

If there is any doubt as to the serviceability of the equipment *do not use it* and seek further advice from a competent person or the manufacturer.

Records should be kept for all PPE and equipment used for Work at Height including the date of purchase, each and every use, all inspections and EC declarations of conformity. Information contained in any thorough examination/ inspection report must also be kept available for inspection.

The *competent person* who carries out such inspections obviously needs special training and instruction that will almost certainly make this person 'competent' to examine many of the items that also require inspection and examination under LOLER. This training must include training to identify the many different types of fault that can occur with the equipment, such as exposure to heat, chemicals, abrasion and shock loadings. At present very little training is available, but some excellent and highly recommended training is available from some of the manufacturers of lifting and fall arrest equipment. This training also includes demonstrations of the test procedures these manufacturers use. In this instance, the manufacturers are best placed to provide such training as they know the products better than anyone and will usually not approve any inspections on their products unless they have trained the inspector.

At present I do not know of many companies who have staff that have undergone such training, but I strongly urge companies to get their staff trained to a 'competent' standard before it's too late.

Lifting equipment used for people or loads, which is subject to **Regulation 9** of LOLER, requires a more detailed comprehensive inspection – called a through examination – which may include some dismantling and/or testing. If this is done it should avoid the need for a more frequent inspection unless the equipment's efficacy depends on how it has been installed or assembled and there have been exceptional occurrences which might jeopardise its efficacy.

However, it is important to remember that some items of equipment for work at height, for example a mast climbing platform, will have some parts which are subject to thorough examination under LOLER. Others, such as floors and guardrails, will not be subject to LOLER and may need to be inspected more often.

Regulation 13 requires that the surface conditions and other permanent features where work at height will be taking place are checked each time before work starts in order to identify whether there are any obvious defects. For example, checking the ground surface on which a tower scaffold or a portable ladder was to be placed. An employer may not be able to do this himself, but should ensure that a competent worker carries out the necessary checks. The results of such checks do not have to be recorded.

Safe Practice

It's far beyond the scope of this book to go into rope access and climbing techniques and the use of fall arrest equipment; these areas require training from specialist trainers and instructors. However, I will say a few words on safe practice.

Never work alone at height. The person who accompanies the worker need not necessarily need to climb as well, but they must know what to do in an emergency and, if necessary, effect a rescue.

Before starting overhead work, the area below should be cleared of anyone who is not essential to the operation. Any objects or items of equipment that a climber may fall onto and that may cause *extra* injury should be moved.

Do not carry or have in pockets anything that could fall whilst you are at height. This includes mobile phones, Mag Lights, coins, keys and Leatherman tools. Provide a suitable receptacle for personnel to deposit the contents of their pockets into *before* going onto the grid.

I have heard of riggers stitching their pockets up so they can't put anything in them. A 10p piece that drops out of a pocket from 50ft could be lethal if it fell on your head, but of course you'll have a helmet on, won't you?

Use a high-sided tray or sufficiently stable bucket to hold work equipment or materials – not a cardboard box!

Use a lanyard to secure tools whilst you are using them. An old film and television studio technique is to secure the lanyard end to a plastic ballcock or even a small plastic buoy from a ship's chandlers and the other to the spanner or tool you are using.

Using this technique, you don't run the risk of finishing the job, standing up, getting tangled and the belt glide on your jeans, to which you have clipped the lanyard, breaking and letting the tool fall!

Try to 'design out' the need to use tools or small parts at height. Use a suitably thick hand rope (in good condition) to raise and lower tools and materials. Use a suitable and correctly rigged pulley if required. (A 'Klein' bag is ideal for pulling up equipment and tools).

If an item is accidentally dropped, then a warning should be shouted such as "*heads!*" Upon hearing this warning, anyone below should take cover and not just stand there looking up!

Those who *by necessity* need to remain, such as groundsmen, must wear a safety helmet. A 'hard hat area' must be established and warning signs erected to this effect. Always warn anyone who may (albeit in conjunction with a system of work) find themselves working beneath you.

Noise should be reduced as much as possible to allow communication between those climbing and those on the ground. A chat to the PA crew would not go amiss, but arrange it so they don't aggravate the PA while you're up the truss.

It is often possible to operate a type of 'Permit to Work' system for overhead work situations, for example, the climber has to obtain permission from the stage manager before he climbs, and this permission is not granted until a 'hard hat' area has been established and the work area made safe.

After the overhead operation, the climber reports back to the stage manager and gives him the 'all clear' to resume normal work. Always advise the relevant personnel on completion of work and that the area is clear of climbers and equipment. If 'house' rules exist (Permits to Work, etc.), adhere to them.

Once a climber has finished his work and has to return to ground level, he/she should do so by the safest possible means. Don't go abseiling off the roof if there are steps you can walk down. Don't try to be macho, you don't impress anyone and you are probably breaking the law by putting yourself at unnecessary extra risk.

Criminal charges will almost certainly be brought against anyone who has consumed drugs or alcohol and who has then contributed towards an accident.

At no time must anyone be forced or bullied into working at height; the choice whether or not to climb always rests with the operative. Never work at height if you are 'off colour', under medication that may affect judgement or reactions, or uncomfortable with the task.

If the C.D.M. Regulations were applied to our work it would mean staging companies would (at the design stage) have to 'build in' safety systems to new stages.

These systems may take the form of traverse wires, wire lifelines on inertia brake drums, and running bars so that climbers and riggers as well as those who build and dismantle temporary stages can attach themselves at all times they are at risk. On existing stages we are required to rig temporary systems for fall protection.

Advanced Access Techniques

Rope Access

Many riggers in our industry are strongly against the work systems employed by IRATA (Industrial Rope Access Training Association) that conform to B.S. 7985: 2002; A Code of Practice for the use of rope access methods for industrial use.

The main objection is towards the use of two ropes, a main rope and a safety backup. Unfortunately, or fortunately (depending on which way you are looking at things), I am certain that systems similar to the IRATA standards are the ones we will soon have to adopt in our industry for the following reasons:

a) The Management of the Health and Safety at Work Regulations require a detailed Risk Assessment and for the risk to be reduced to the lowest possible level, the risk assessment will obviously show that a second rope will reduce the risk by a further 50%.

I suspect that many riggers have not produced risk assessments that are 'suitable and sufficient'; otherwise this would be an obvious fact.

b) The PPE regulations require the use of PPE to be 'suitable and sufficient' to establish a safe system of work, the use of a second rope is considered suitable and sufficient use of PPE.

c) LOLER requires carriers of people (harnesses) to be fitted with devices to prevent free fall, which should be independent of the primary suspension means, i.e. a secondary safety rope when abseiling or prussicking!

The Work at Height Regulations require two ropes to be used for rope access unless it can be proven (by risk assessment) that one rope would be safer. There may be the odd occasion when it is possible to prove that one rope is safer than two but these occasions will be few and far between and appropriate measures must still be put in place to ensure safety. The HSE recognises that some industries have genuine concerns about the application of single and double rope requirements and are looking at what further guidance would be appropriate for these situations.

The IRATA Code of Practice has been examined by the HSE and in a court of law after an accident and was considered to be a safe working practice, so it is highly unlikely that we can get any dispensation with standards like this having already been set.

At this time I am not advocating that all riggers under take IRATA training. As it stands, IRATA training does not meet all of our industry needs, and only covers rope access and not fall arrest.

We need a work system and the relevant training that covers both rope access, climbing and fall arrest systems.

The way forward is perhaps working with IRATA and the fall arrest training companies to develop industry bespoke systems together with training that meets our needs and meets all the other legal and safety criteria.

At the time of writing a new national training and certification scheme for riggers is about to be launched in the UK that includes a work at height module. Let's hope this fulfils the industry needs and that the work at height module will be available as a 'stand alone' module so that the other trades (not riggers) can take advantage of the training and certification.

For anyone working at height in a roof space, the period of greatest risk is the traverse to the workstation. This has been the subject of concern for some time.

The worker can progress horizontally, either by 'leapfrogging' lanyards, or by using an appropriate horizontal safety system. Horizontal systems (temporary and permanent) can be installed in roof spaces and can provide a high level of worker protection at all times the worker is 'aloft' and connected to it. (The systems are usually not applicable to use on lighting trusses).

The use of such systems requires special training and a structural engineer to calculate the forces imposed by such devices on the structure to which they are attached; however, this is beyond the scope of this book.

What Kind of Rope?

Selecting the correct type of rope for the job is vitally important. Basically, ropes can be split into three main categories, but only two are of interest to us in a work situation:

a) Dynamic rope: this is very high stretch rope, almost elastic-like. It is designed to absorb the impact of a falling climber and is normally made from nylon with a diameter of 11mm. These ropes are normally constructed from very long filaments of nylon (the kern) with an outer (often brightly coloured) braided sheath or mantle. They are often known as kernmantle ropes. Because of their high stretch characteristics they are not recommended for ascending (prussicking) or for frequent abseiling (descending); check for sheath slippage if they are used for abseiling.

These are specialist ropes; don't use them as towropes or for equipment lifting. They are sometimes available with an 'Everdry' finish that helps prevent water absorption and aids wear resistance.

The shock load is the force which is transmitted to the climber, karabiner and all components of a belay as the consequence of a fall. Its value is the single most important factor in a climber's security. Lowering the shock load results in:

- Reduction of the force on the climber as the fall is arrested
- Reduction in the risk of belay and anchor weakness or failure, due to the reduction in the force they sustain
- Reduction in the dynamic breaking distance due to lower forces at the belay

Dynamic ropes are mainly for leisure and recreation purposes such as rock climbing. They may not meet industrial standards and should not be purchased for use at work in the entertainment industry.

b) Static rope: this is low stretch rope and has been designed for cavers using Single Rope Techniques (S.R.T.). Also know as speleo rope or abseil/prussick rope, they are designed to absorb only the minor falls of a climber as they only stretch about 6%. In fact, a serious fall could easily break such a rope or halt a climber's fall so rapidly that it breaks every bone in the climber's body. They are, however, made to industrial standards and when correctly used they are suitable for fall arrest. For this reason they are the only type of rope to be considered for use in industry and the only type of rope that should be purchased for our work. These ropes are made of nylon and are normally 11mm in diameter. Like dynamic rope they will normally have an inner core with one or even two outer sheaths that are usually designed not to slip.

These are specialist ropes; they are not to be used as tow-ropes or for equipment lifting. Since they are of such small diameter, they can be dangerous to grip with heavy loads for any period of time. Usually the word 'Speleo' will be found on a label or markings for these ropes.

'Everdry' finishes are often also available but one of the most important factors of these ropes (apart from low stretch) is their very high abrasion resistance, making them doubly ideal for use with mechanical ascending and descending devices and for 'self life-lining'.

Care must always be taken when abseiling, and smooth, slow descents should be made to prevent the build-up of heat in mechanical descenders which could melt nylon ropes.

These ropes must be individually identifiable, be clearly marked with its Safe Working Load and have a CE mark and EN 1891: 1997.

Ropes purchased new and ready cut to length from an industrial supplier will carry this mark. If a drum of rope is purchased from an industrial supplier, a Certificate of Conformity will be supplied. The user can then cut rope off the drum to the required length and mark it to show that it came from the batch off the drum and therefore conforms.

For lifting use or for rope access, use of the marking is essential to comply with the Lifting Operations and Lifting Equipment Regulations 1998.

Safety Chains (by Chris Higgs)

We are all familiar with the 'safety chain', but would it 'hold' a 2kW lantern when the lantern falls the full length of the chain? It is commonplace to see 2 and 3mm wire rope 'safety bonds' being used as secondaries for items such as speaker cabinets weighing 16kgs. *They will not work!*

A single 6mm wire rope with a suitable connector (shackle, maillon or possible karabiner of 'rated' strength) may hold a load of 398kgs, *but only if it is tight to start with*, used correctly (no small radii, no sharp edges, used at 0 degrees, etc.) and fitted to a suitable point above and on the equipment.

The use of inadequately sized steel wire bonds, looped/knotted chain bonds and Crue hooks is potentially hazardous.

Amendment 3 1993 to British Standard BS 4533 102 .17 specifies:

> "...with the end of the secondary suspension remote from the luminaire securely fixed, the luminaire is allowed to hang freely by means of the secondary suspension alone.
>
> The luminaire is raised in its hanging mode a distance of 300mm vertically and allowed to free fall. This test is made 30 times.

The secondary suspension shall not fail and no part of the luminaire shall fail."

Any safety bond used should be sized according to the weight of the equipment it is being used for and connected to a structure capable of holding the force generated by the equipment falling freely.

Using the bond 'doubled' (so the material and connector take only half the force of the fall) will enhance the strength of the bond. The connector should be of appropriate design and strength, and be subject to regular inspection and maintenance.

Using wire rope is preferred because it is stronger in use (over small radii and hard materials) than chain, and its stiffness means it will not interfere with the operation of the luminaire.

N.B.

Individual heavy objects (video projectors, 'intelligent' lanterns etc.) should be regarded with caution. Even with a good 'safety' the shock load of the item falling could cause structural problems at the point where it is fixed, particularly on a span of truss or barrel.

32 LIFTING OPERATIONS AND LIFTING EQUIPMENT REGULATIONS 1998 (LOLER)

Chris Hannam and Chris Higgs

These regulations are an amendment to the Provision And Use of Work Equipment Regulations 1998. The regulations came into force on 5th December 1998, and on that date many of the existing lifting regulations were revoked. The LOLER regulations cross-refer to PUWER 98. Duty holders now have to comply with all the requirements of the regulations. LOLER covers all lifting equipment; this includes existing equipment and second-hand or leased equipment. Regulation 2(1) defines 'lifting equipment' as "work equipment for lifting or lowering loads and includes its attachments used for anchoring, fixing or supporting it". It includes any lifting accessories that attach the load to the machine in addition to the equipment that carries out the actual lifting function. The scope of the Regulations is therefore very wide and includes a range of equipment from an eyebolt to a tower crane.

Regulation 2(1) defines a 'load' as any material, people or animals (or any combination of these), that are lifted by lifting equipment. In some circumstances, the weight of the lifting accessories will need to be considered as part of the load being lifted.

The Regulations are primarily aimed at the type of equipment that was covered by earlier lifting legislation, i.e. cranes, lifts, hoists, and components including chains, ropes, slings, hooks, shackles and eyebolts. LOLER also applies to a range of other lifting equipment that presents risks, similar to those associated with the traditional equipment listed above such as harnesses and hardware used for suspended rope access and work positioning. It does not cover equipment used for fall arrest; instead, this is covered by the PPE and Work at Height Regulations. However, it must be noted that on occasions the same equipment is used for both rope access/work positioning and fall arrest. In this situation, it must comply with LOLER. The regulations certainly apply to forklift trucks, cherry pickers, scissor platforms, trussing, round slings, rigging motors, shackles, steels, in fact all the lifting equipment we regularly use in our industry.

Another interesting point is Regulation 5, which says that carriers (harnesses) used to lift people should be fitted with devices to prevent free fall, which should be independent of the primary suspension means. In other words, a secondary safety rope must be used when abseiling or prussicking.

It is important to understand that these regulations were introduced as part of European attempts to harmonise regulations between member countries. Therefore, the HSE now has the task of introducing LOLER as it is part of the regulations needed to implement the 'Amended Provision and Use of Work Equipment Directive'.

The regulations are 'goal-setting' but do not introduce many changes from current best practice and the old regulations. In case you don't know, it's been a requirement for 'competent' persons to examine, operate and supervise for some 40 years and the requirement to carry out risk assessments has been in force since 1992 (and arguably since 1974).

The HSE has produced an Approved Code of Practice for LOLER (and many of the other regulations under HASAWA) and this will be considered as the model of best practice in any court proceedings.

If an employer can't demonstrate compliance with the ACOP, he will almost certainly be found guilty of any breach of the regulations unless he can produce evidence that he had alternative systems in place that exceeded the standards set by the ACOP to achieve a higher standard, and then produce evidence of that achievement. It is the author's view that it will be extremely difficult, if not impossible, to match or attain higher control standards than those set by the ACOP.

Key Points of LOLER

1. Any 'lifting equipment' or 'accessory for lifting' used to lift or position a piece of work equipment will need to have an inspection/examination carried out by a 'competent person'. This first assessment will take into account how the equipment will be used as well as how often it will be used as well as the environment in which it will be used. The information gained as a result of this initial assessment is then used to make the choice between an inspection/examination/test regime designed by a competent person or inspections/examinations/tests at minimum mandatory intervals. The mandatory intervals for test/inspection/examination are 12 months for most items but every six

months for equipment that will lift or support people (such as rope access equipment). The initial assessment will decide if examination/inspections/tests are required more often than those that are mandatory.

In addition, LOLER requires a thorough examination to be made following any significant change which may affect the safe operation of the lifting equipment (such as a sudden shock loading). Written risk assessments must also be made to determine if more frequent inspections are required in addition to those already listed.

LOLER requires records to be kept for all lifting, work positioning and rope access equipment, including the date of purchase, each and every use and all inspections; this is a legal requirement. It is obvious that to make such a record all equipment must be individually identifiable and clearly marked. This detailed marking is also a requirement under L.O.L.E.R. Items purchased in bulk and stocked in large quantities, such as shackles, may be batch marked, e.g. by spraying each batch a different colour and recording the details of each batch and its colour code. Records of inspection/examination/test must be kept until the next inspection/examination/test is made. EC declarations of conformity must be kept by an employer for as long as he or she owns the equipment.

2. The 'competent person' for inspections of equipment may (or may not) be someone in the employment of the owner of the equipment. In-house examiners must have the genuine authority and independence to ensure that examinations are properly carried out and that the necessary recommendations arising from them are made without fear or favour.

It has been suggested that to pass the 'test' of independence, the 'competent person' will be required to have full authority, without interference by line managers or owners, to scrap, modify or repair equipment as required. The 'competent person' who makes inspections also has a duty to inform the relevant enforcing authority where there is in his opinion a defect in the lifting equipment involving an existing or imminent risk of serious personal injury.

3. All risk assessments, testing and inspection will have to be recorded and employers will need to ensure that no lifting equipment leaves his or her undertaking without physical evidence (documentation), and that the last thorough examination required has been carried out.

'Documentation' means any retrievable recording system.

It has been claimed that the HSE has stated that there will be no requirement to send out all the documentation with the equipment, provided it is available for inspection at reasonable notice. The Statutory Instruments for LOLER (the Law) state that the documentation *must* accompany the equipment and a capable lawyer could demonstrate, in court, a strong case against you if you use the excuse: "the HSE said you don't need to carry the documentation with the equipment", as there is nothing to support this claim.

Regulation 9(4) of the regulations states that the information accompanying the equipment needs to include:

a) The name and address of the duty holder for whom the thorough examination was made:
b) The address of the premises at which the thorough examination was made
c) Sufficient information to identify the equipment.
d) The date of the last thorough examination.
e) The date when the next thorough examination is due.
f) The safe working load of the equipment or, where safe working load depends on the configuration of the equipment, its safe working load for each configuration of equipment.

4. It is quite clear that many items currently not inspected by our industry will now have to be inspected. Very little is new, however, as most of these items, perhaps unaware to you, probably needed inspection and examination under previous regulations. In a nutshell, only the actual load (i.e. luminaire, loudspeaker, projector, etc.) will escape and even then each load item should be marked with its weight (also, see the section on Safety Chains in the chapter on Work at Height).

5. Every lifting device or accessory has to be clearly marked with its Safe Working or Safe Load Limit (Regulation 7a). Any equipment that can be assembled in a number of ways with different loading characteristics (such as truss) must be accompanied by sufficient information for the user to assemble and use it correctly (Regulation 7b). Documentation regarding the SWL or SLL of its configuration will have to be available on-site.

These are known as essential Health and Safety requirements and are required under the Machinery Directive of the Provision and Use of Work Equipment Regulations 1998.

All aluminium truss systems should conform to B.S. 8118 (Structural Use of Aluminium) and steel components to B.S. 449 (Structural Use of Steel) [LOLER Regulation 4].

The manufacturers of CE marked equipment are required to provide the purchaser of new equipment with information on the correct use, care, maintenance and inspection of marked equipment as well as details of testing. It may be that a similar situation applies to modular fly-bar systems used for P.A. systems and video screens. The person who assembles the equipment will also need to be 'competent'.

The Approved Code of Practice for LOLER requires that where the weight of a Lifting Accessory is significant in relation to the Safe Working Load of the Lifting Machine with which it is used, the accessory should be marked with its self weight.

Again, this applies to trusses, PA fly-beams, etc. which can be a significant proportion of the capacity of the hoists used to lift them. An individual shackle is not a significant proportion of the capacity of such a hoist and so would not need to be marked with its self weight.

6. All equipment used in the EU is now covered by the new regulations. Consequently, UK companies may for once be at an advantage over American companies. LOLER could work in our favour, particularly with equipment that is CE marked.

7. Any person, either employed or self-employed, who 'has control' of lifting equipment comes under LOLER just as much as the employer. 'Control' refers to anyone (such as a rental supplier, consultant, competent person, specifier or supervisor) who can effectively control the way in which equipment is used *in terms of safe use on site.* It has nothing whatsoever to do with the operating of a controller.

Repetitive operations need only be assessed once, but as with all assessments, monitoring should take place and reviews undertaken to ensure that the Lifting Operation and the assessment has not changed in any way. One idea is to classify work as routine and non-routine.

Employers have a responsibility to ensure that anyone they ask to carry out lifting operations is suitably competent. The person responsible for the lifting operation will have to ensure (and be able to demonstrate) that the lifting operation is:

a) properly planned by a competent person
b) supervised by a competent person
c) carried out in a safe manner

Any employer (such as a hire company, promoter or even a production

manager (if the production manager is the hirer) is obliged to ensure that anyone they hire/employ to be in control of lifting equipment is competent.
8. Regulation 8(1) states that loads should not be moved by the use of Lifting Equipment when there are people underneath, unless adequate control systems are in place to protect them.

A stage therefore needs to be cleared while trusses and other equipment are raised and lowered during a load in or load out for a show. However, it will still be possible to use moving trusses during a show, provided adequate systems are in place that will fail safe.

9. After installation, all rigging (for lighting, video, PA, etc) and its fixing points, must be inspected and checked by a 'competent person' to see that it is safe. This 'competent person' could issue a certificate following the installation inspection, something which is considered good practice.

This is where training and qualification are so important: without training and qualification, how else are you going to prove 'competence'?

The Chain of Responsibility

The employer of persons carrying out Lifting Operations holds ultimate responsibility for compliance with the regulations. A production manager, for instance, who hires a rigger will hold the responsibility, but if the production manager is *employed* by a promoter or artist management, then the promoter or artist management will also take on the responsibility. Employers also hold responsibility for any Lifting Equipment used in any Lifting Operation.

The responsibility of the employer cannot be passed on or delegated. Everyone in the chain of an employer/employee relationship may be held liable in the event of prosecution or action by the enforcement authorities.

A band, artist management company or promoter who employs a production manager, who in turn employees a rigger (even on behalf of the band, management company or promoter) could be liable for compliance with the regulations. This responsibility cannot be avoided by employing persons who own their own lifting equipment.

The Paper Trail

As already stated, 'Documentation' means any retrievable recording system, keeping records and being able to produce them. Documentation forms a major part of compliance with LOLER as it does with many Health and Safety regulations.

a) If a Lifting Operation is entirely 'in house', no documentation needs to be given to a third party. 'In house' means that the company own the lifting equipment, employ the staff and control the Lifting Operation. It does not matter where the operation takes place. In this situation, the 'documentation' will only need to be available for inspection for the enforcement authorities.
b) Lifting Equipment that is not used 'in house' will need to be accompanied by its documentation.
Persons who hire (or borrow) lifting equipment from someone else have the legal responsibility to ensure that LOLER is being complied with. To carry out this responsibility, they must be in possession of the relevant paperwork to prove that it has been examined and/or inspected.
The paperwork must accompany the equipment at all times that it is being used by a third party and away from 'base' location.
c) As stated, LOLER cross-refers to the Provision and Use of Work Equipment Regulations 1998 and the Management of Health and Safety at Work Regulations 1999. The management regulations require a risk assessment to be carried out to identify the nature and level of risks from Lifting Operations (and any other hazard).

Every individual (self-employed), partnership or company that is responsible for Lifting Operations (or has employees that carry out Lifting Operations), is responsible for making sure that risk assessments have been carried out and are available for inspection by the enforcement authorities.

LOLER at a Glance

If you are a promoter, event organiser, production manager (assuming the production manager is the employer), artist or artist manager, and you hire riggers, rigging equipment or stage roofs for your shows, then you should ensure that all the contractors you use are complying with LOLER and all other regulations. If you hire riggers or hire stage roofs, you will have the same liabilities as shown below.

Promoters and production managers may be liable for any Lifting Equipment used on a show they promote or produce. This could have an impact on productions from outside the EU where no equivalent to the LOLER exists.

You should:

a) Ensure that all persons you employ to carry out Lifting Operations are 'competent' and can prove this 'competence' by documented evidence.

b) Ensure that all Lifting Equipment (and equipment used to support Lifting Equipment such as stage roofs) has adequate LOLER documentation accompanying it. Keep the documentation with the hired equipment.

c) Carry out (or have a 'competent person' carry out) risk assessments for all Lifting Operations and have these assessments available for inspection when required by the enforcement authorities.

d) Make certain that Safe Systems of Work are in use (e.g. don't lift or lower loads such as trusses when people are underneath).

Self-employed Riggers (self-employed persons are treated as employers under the HASAWA)

You have the same legal responsibilities as anyone else who uses Lifting Equipment. You have to make certain that you are 'competent' and you should ensure that anyone you provide your services for is complying with LOLER. In particular:

a) Ensure all Lifting Equipment you are required to rig has documented evidence of examination/test/inspection.

b) Ensure that your own equipment used for work positioning (harnesses, slings, karabiners, ropes, etc.) is individually identifiable, marked with its SWL or SLL, has been inspected/thoroughly examined and risk assessed by a 'competent person' to decide if an inspection/examination regime by a competent person is required or if mandatory inspections/thorough examinations (by a competent person) are adequate. This initial risk assessment will consider the frequency of use of the equipment, the nature of the work and the work environment.

The risk assessment by the 'competent person' will almost certainly find that additional inspections are required inbetween the mandatory inspection/examination that must be carried out every 12 months, or every six months for equipment used to lift people. Records of all tests/inspections and examinations must be recorded in writing and details kept with the equipment.

Finally, you the user should inspect your equipment each day before you use it.

Rigging and other service companies owning lifting equipment must appoint a 'competent' person to risk assess the use of the lifting equipment and decide upon the frequency and system for inspections/through examinations of the equipment.

a) You must make certain that all lifting equipment is individually 'identifiable' or 'identifiable' by batch.
b) You must ensure all equipment is marked with its SWL or SLL.
c) You must have the equipment inspected and thoroughly examined by a 'competent person' and the results recorded.
d) You must supply evidence of the inspection/thorough examination to any third party using the equipment.
e) You must get LOLER documentation from the owner/hire company of lifting equipment you hire; this documentation must be kept with the equipment.
f) You must appoint a 'competent' person/s to plan, assess and supervise all lifting operations; a copy of the assessment must be available on-site for inspection.

Under LOLER, any venue with designed lifting points ('house points') in the roof, or any kind of structural lifting grid will have to have these inspected and examined by a 'competent person'. The inspection and examination reports must be available to all users of the venue who use these 'house points'. By doing this the venue will have complied with LOLER. This has been a legal requirement at least since the Factories Act 1961.

Local authority structural surveys for Entertainment Licence purposes are carried out every three years, and the various 'old' regulations required lifting points to be examined and recorded in the same way LOLER does.

LOLER Definitions

Competent Person (to carry out inspections and thorough examinations or to assess Lifting Operations)

A person aware of the limitations of their own experience and knowledge, who understands current best practice, who is willing and able to supplement existing experience and knowledge, who has access to specific applied knowledge and skills of appropriately qualified specialists and who has the appropriate qualification where such a qualification exists.

Lifting Accessory

Any individual items of lifting equipment such as shackles, span sets, eyebolts, sections of truss, hook clamps, etc. which cannot provide motion by themselves.

Lifting Assembly

A collection of Lifting Accessories used to join a load (or combination of loads) to Lifting Equipment or to support the Lifting Equipment itself. Assembled trusses, ground supports, and stage roofs all fall under this category.

Lifting Equipment

The generic term for all items found in the workplace that can be used to lift a load and that provide motive force, manual or powered.

A few examples of power-driven machines include rigging motors, forklift trucks and cranes. Examples of manual machines include Tirfor winches and manual chain hoists.

To Sum Up

LOLER is exactly what the entertainment industry needs. For the cowboys who have not had their equipment inspected or examined (don't forget, this was a legal requirement before LOLER) it may come as a shock.

For the responsible companies who have had shackles, slings, hoists and so on examined and, where necessary, tested (hoists particularly, because of the need to assure ourselves of their mechanical integrity in an environment with no room for doubt over safety, ethically or financially), there should be no real changes unless they wish *to change the frequency of their inspections and examinations, because they can now rely more on risk assessment than regulations.*

The only real changes are the need to include as lifting 'accessories':

1) Trusses, PA beams, and so on.
2) Genuine rope access or load bearing personnel harnesses and associated suspension equipment, such as slings and karabiners. (This is a situation where equipment normally classed as PPE can be classed as Work Equipment and is therefore covered by LOLER).
3) Items used to lift people as 'carriers' (skips, cages, or baskets, etc.) Employers must carry out risk assessments; this has been a requirement since at least1992.

All lifting equipment used in construction, factories and in most theatre uses, has been covered by regulations under the Factories Act (Construction Regulation series).

Competent persons have been required to examine equipment, operate equipment and supervise for at least 40 years. If you think LOLER regulations are all new then you must have been asleep until now.

We can now take control of our own destiny. This is exactly what we have complained about for years - being told we have to do things in a certain way. From a safety/engineering standpoint, we should use the same standards and criteria as the engineering world – generally speaking, we are using the same equipment for very similar purposes.

This is our chance to prove we are capable of setting our own standards, using our own code of practice and almost police our own industry in terms of what we expect of contractors and suppliers, workers and co-workers.

If we cannot be bothered to accept a minimum standard of safety and accountability, then we must be as guilty as the cowboy promoters, management and service companies we all complain about so much.

LOLER FLOWCHART

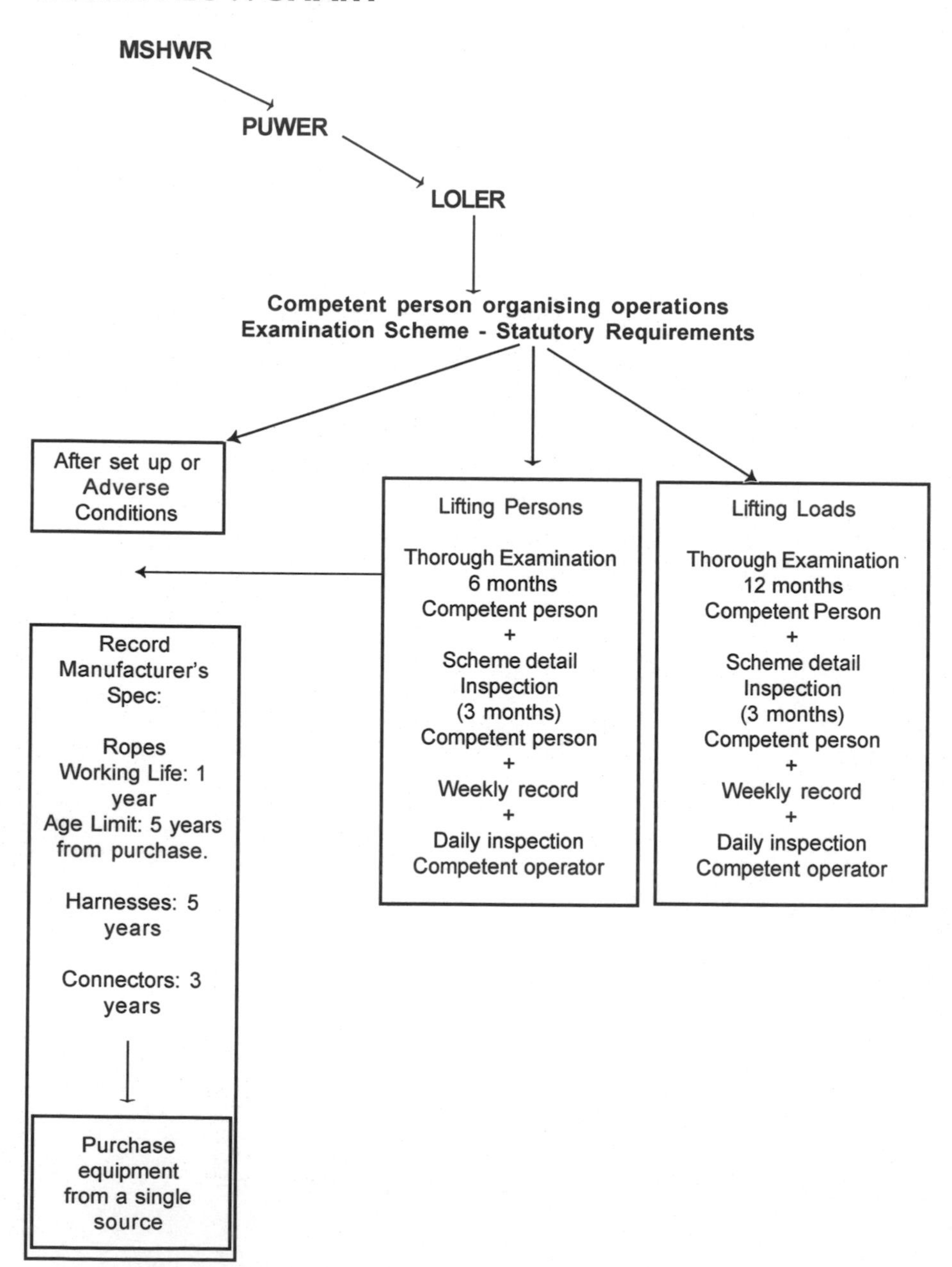

THE LIFTING OPERATIONS AND LIFTING EQUIPMENT REGULATIONS 1998

LOLER 1998 details requirements for organising operations, the selection of work equipment and the mandatory inspection regime.

As such it classifies the equipment as 'Work Equipment' rather than 'Personal Protective Equipment'. The equipment is therefore also covered by the Provision and Use of Work Equipment Regulations1998 (PUWER).

See Key points chart on following pages.

TOPIC	LOLER KEY POINTS	Reg. No
Organisation	Risk assessment will identify the hazards and corresponding risks. Employers to ensure every lifting operation is: · Properly planned by a competent person. · Appropriately supervised. · Carried out in a safe manner. The suitability of equipment (or combinations) and avoidance of having loads over people are detailed. Also ensuring that potential danger zones are not entered. Relevant issues in lifting: · Visibility, attachment, environment, location, overturning · Proximity hazards, de-rating, lifting persons, overload · Pre-use check, continual integrity.	 8(1)a 8(1)b 8(1)a-c
Suitability of Equipment	Detail on selection of suitability in construction and ergonomic risk. · Material of manufacture · Means of access · Protection against slips, trips and falls · Operator protection · Effects of high wind.	3
Strength and Stability	Adequate strength for use. Adequate stability-destabilising forces. · Equipment not susceptible to foreseeable failure. · Appropriate safety factor against failure used. · Suitable effective measures to prevent overturning. · Ground/aerial factors that can affect mobile equipment. · Requirements for equipment on rails. · Provision of tyre pressure gauge for mobile equipment. · Use of rated capavity indicators - audio/visual devices. Every part of a load/anything attached to it for lifting is adequate.	4(a) 4(b)
Lifting Equipment used for lifting persons	Avoidance of risk of crush, trapping, being stuck or falling while in or working from the carrier. Persons trapped can be freed. · Provision for use of non purpose-made carriers (forklift). · Applications to lift cars. · Edge protection and non-slip floor-falls over two metres. · Independent means of suspension - e.g. check valves. · Reliable means of rescue in event of malfunction.	 5(1)a 5(1)b 5(1)c 5(1)d
Marking of lifting equipment	· Equipment and accessories for lifting loads marked with SWL. · Machinery marked with SWL for each configuration with machine. · Accessories marked to show characteristics for safe use. · Equipment for lifting persons is clearly marked to show this. · Equipment not meant for lifting persons is clearly marked so.	7(1)a 7(1)b 7(1)c 7(1)d 7(1)e

Thorough Examination (TE) and Inspection	All inpection programmes are set after risk assessment by a 'competent person'. This is called the Examination Scheme. It details the frequency, type and competence required. Mandatory TE frequency is set as a minimum. · TE before use unless: It has not been used before and has CE conformity, issued not more than 12 months before. Or, it is accompanied by physical evidence of a current TE. · TE after installation or re-assembly on a different site. · TE if deterioration will result in dangerous situations. Lifting persons - six months Lifting loads - 12 months After any exceptional circumstance in accordance with the Examination Scheme, including 'in-service' Inspections by a competent person. · Physical evidence of current TE accompanying equipment when leaving an employer's undertaking. If obtained from another undertaking and used as yours, the competent person completing the TE is usually external to the organisation. Internal inspectors can be used providing there is no chance of shoring fear or favour. They should have appropriate practical and theoretical knowledge and experience of the equipment to be examined.	9(1)a 9(1)b 9(2)ab 9(3)a 9(3)ab 9(4)a 9(4)b
Reports and Defects	Duties of the Competent Person completing the TE: · Report any defect which is or could be a danger. · Make a report as soon as is practicable, authenticated, containing the information in Schedule 1. · Report defects risking serious or imminent danger to the Enforcing Authority. · Inspection reports to be made to the employer in writing as soon as is practicable. <u>Not TE report.</u> · Employers notified to ensure defective equipment is not used and meet any recommended time-scales.	10(1)a 10(1)b 10(1)c 10(2)ab 10(3)ab
Keeping Information	Duties for keeping information: · CE declaration of conformity to be kept as long as the equipment is operated. · Employer to keep information from TE for <u>lifting equipment</u> until he ceases to use it. · Keep TE for lifting <u>accessory</u> for two years after the report. · Installatiuon TE report kept until equipment disassembled. · Inspection reports kept until next inspection (or two years)	11(1) 11(2)a i 11(2)a ii 11(2)1 iii 11(2)b

Quote: "Accidents happen because of lack of time, poor communications or bravado.

Who ever invented the phrase "the show must go on" did not give a damn about the crew who put the show together, and without enough time and little sleep they are the ones at risk." ***Bill Sapsis. Sapsis, Rigging Inc. USA.***

33 SPECIAL EFFECTS, LASERS, STROBES, UV LIGHTING, SMOKE, VAPOUR EFFECTS AND PYROTECHNICS

Information on the use of strobes, UV lighting, smoke, vapour effects and pyrotechnics may also be found in the 'Pop Code' and 'Fire Precautions in Places of Entertainment', etc. The manufacturers' instructions should always be followed. As with all hazards, the risks from these effects need to be assessed and then controlled.

An Entertainment Licence granted under the Licensing Act may contain conditions that ban or restrict the use of fireworks, pyrotechnics, strobes, smoke and vapour effects, lasers, UV lighting and many other kinds of special effect.

Outdoor events that intend to use fireworks, lasers, searchlights or high power lighting effects should consider the dangers these effects can have on aircraft pilots or the aircraft themselves, especially if the event is to be held within a twenty mile radius of an aerodrome. The Civil Aviation Authority has produced a Guide for the Operation of Searchlight, Lasers and Fireworks in United Kingdom Airspace. This guide also contains a simple form to notify the CAA of any proposed event using these effects. Production managers, display operators and event organisers should obtain and follow this guidance, or they may risk prosecution.

Strobe Lights

Strobe lights are known to induce epilepsy in some flicker-sensitive people; this is a rare condition but if an episode is triggered, there is a high risk that the flicker-sensitive individual will experience a full seizure.

The advice for strobe lights is that they should operate on a flicker rate of 4 or fewer flashes per second (as then only 5% of the flicker sensitive population will be at risk of an attack). This flicker rate only applies to the overall output of any group of lights in direct view, but where more than one strobe light is used, the flashes should be synchronised. The same guidance applies to intelligent lighting systems that are capable of producing strobe lighting effects.

Promoters are advised to warn the public by signs, warnings in the programme, or announcements over the PA system if strobe lights are to be used as part of the performance. First Aiders and stewards trained to deal with epileptic fits should be on hand and staff should also be given warnings.

Ultraviolet Lights

In some very rare cases some people have become photosensitised by ultraviolet lighting on exposed parts of the body. These people may have increased sensitivity to the effects of UV light on the skin; this can be because of one or more medical conditions or because they are taking medication of some kind.

The result of such sensitisation may be result an exaggerated sunburn reaction or other skin changes. Further exposure should be stopped and that person should seek medical help.

UV lamps should always be used, repaired and bulbs replaced in accordance with the manufacturer's instructions. They should also be kept as far away from people as possible to restrict the exposure to ultraviolet radiation. Performers and crew should be made aware of the risk of UV radiation and warning signs erected backstage.

In order to remove UVB radiation, some lamps have a double skin whereas others have lamp housings with a separate special filter; lamps should not be used if this filter is missing or the outer skin is broken.

Smoke, Vapour and Fog Effects

There are four main ways of generating smoke and vapour:

- Liquid Nitrogen or solid Carbon Dioxide (also known as dry ice)
- Cracked oil or cracked water
- Fluid based smoke machines
- Pyrotechnic smoke effects

Pyrotechnic smoke devices often give off harmful by-products and are not to be recommended for use indoors, especially as so many other safer methods of generating smoke and vapour are available.

Mineral Oil

This includes oil crackers and diffusion foggers. Oil crackers involve bubbling air through a drum of mineral oil.

The air bubbles reaching the surface contain 'cracked' oil of particle size 1 to 50 microns. This oil is not 'cracked' in the sense of chemically breaking down the oil but is merely creating smaller droplet size. This has also been used in combination with dry ice.

The diffusion fogger produces a mineral mist of less than 1 micron size by using a compressor to force mineral oil through a series of fine filters.

No toxicological studies have been made on inhalation of mineral oil of particle size less than one micron. However, there is concern about long-term problems such as lipid pneumonia, since the very fine mineral oil mist gets deep into the lungs and stays there. This is not recommended for use indoors or where personnel could inhale the mineral oil.

Glycol Fogs

During the last decade, a whole range of products has been developed that uses mixtures of water and polyfunctional alcohols, including ethylene glycol, propylene glycol, diethylene glycol, triethylene glycol, polyethylene glycol and glycerine. In general, these are safer than most of the other fogs and smokes with the exception of dry ice. Ethylene glycol and diethylene glycol are toxic by ingestion, causing kidney damage and possible death; the other glycols mentioned are considered only slightly toxic.

Ethylene glycol has been removed from most fogs after studies showed that it is a teratogen (can cause birth defects).

Unfortunately, long-term studies have not been done on inhalation of the mists of most of these glycols, although respiratory irritation is sometimes listed on Safety Data Sheets.

A more serious concern is how the fog is generated. These mixtures are heated in a fog machine that heats the liquid to a temperature nearly 600F. One air sampling study found significant levels of acrolein in the mist generated.

Acrolein is a strong respiratory and eye irritant.

It is likely that some chemicals could generate more decomposition products than others.

Reformulation and finding ways to reduce the temperature needed to create the mist are possible solutions. Despite these problems, at this time the glycol fogs are probably the least hazardous fogs to use, although some will probably turn out to be safer than others.

Some basic safety rules apply to all types of smoke machine:

- The machine must be in a fixed position and adequately protected from interference, as some machines can get very hot during use

- A competent operator should be with the machine
- All machines must be used in accordance with the manufacturer's instructions.
- Minimise the exposure time to smoke and vapour for all those concerned
- People should be kept away from areas in front of all machines since concentrations are at their highest here.

Fans may be used to direct the smoke or vapour but care must be taken to prevent the spread of smoke or vapour into public areas, since this may cause an audience to panic. Tests should be carried out prior to the performance.

Smoke and vapour must not be discharged or allowed to drift into exits, stairways, escape routes, or be allowed to obscure exit signs or fire protection equipment. Some fluids, cracked oil in particular, leave a deposit on stage which can prove a dangerous slip hazard: spillages of oil or fluids can also cause serious slip hazards and should therefore be cleaned up immediately.

If fire or smoke detection equipment is fitted within a venue, special care needs to be taken. Some venues and Entertainment Licence conditions do not allow the use of smoke or fog machines because of these detectors.

If fire or smoke detectors are in use, the amount of smoke or fog must be restricted to prevent these detectors operating. On no account try to prevent smoke detectors operating by covering them, overriding or switching them off.

The majority of water-based smoke fluids and the smoke vapours they produce do not carry a Hazard Rating under the C.O.S.H.H. Regulations. Therefore, an assessment is not required, but it is still necessary to obtain the Safety Data Sheet from the supplier or manufacturer and check this is the case with the fluid you are using. The CHIP 2 (Chemicals (Hazard, Information and Packaging) Regulations may also apply to smoke fluids; users of these fluids need to do some research into this area.

Some smoke fluids are known to be carcinogenic and should be avoided, but fortunately, these are quite rare these days. Some smoke effects are known to cause discomfort to those who suffer from smoke allergy or asthma and these later conditions are probably psychosomatic with modern fluids. Some people may suffer from skin irritation from the mineral oils or glycol used in smoke fluids.

The results of assessments will show if these substances are toxic or flammable and that:

- Good ventilation is necessary when using dry ice or liquid nitrogen, since, after the initial generation, the vapours become invisible, and as both carbon dioxide and nitrogen gases are asphyxiants, concentrations can present dangers to the audience, artists and crew.

- Special care needs to be taken at indoor venues that have orchestra pits and under-stage basements, as these gases may flow into these areas through openings, floorboards and cracks, thus putting people there at risk.
- Dry ice and liquid nitrogen should be stored in the containers in which they were supplied and must only be handled when wearing long insulated gloves and a face visor to avoid splashes, as even momentary skin contact can cause frostbite and blisters. The use of tongs should be considered.
- Some young people and people with asthma or breathing problems find it difficult to cope with certain types of smoke and vapour effects. So, alternatives should be tried.

As productions have no control over the composition of their audiences, special care needs to be taken to minimise risks to which they may be exposed. Ideally, audience exposure to effects should be avoided altogether. Operators and production managers should know if smoke is likely to reach the audience and that if it is likely, the amount of smoke/vapour must be restricted to the minimum amount necessary for the desired effect. As with strobe lights, promoters are advised to warn the public through the use of signs, warnings in the programme, on tickets or announcements over the PA system if smoke effects are to be used as part of the performance.

The HSE has published guidelines for the quantities of smoke, oil and dry ice in the atmosphere (Occupational Exposure Standards) to which people can be exposed.

Unfortunately, a Parliamentary Bill (the Fireworks Bill) did not get through Parliament and become legislation. If it had, it would have 'tidied up', and in some cases replaced some of the existing regulations as well as giving some better definitions to some grey areas of the law. This in turn would have clarified the whole area of firework and pyrotechnic legislation – a situation that was considered by the industry to be long overdue.

In September 2003, the Fireworks Act received Royal Assent. This new act will give government the power to:

- Impose a noise limit of 120dB on fireworks available to the public.
- Ban the use of fireworks during anti-social hours.

- License people who sell fireworks.
- Allow Local authorities to refuse or revoke licences if retailers act inappropriately, and
- Create a compulsory training course for operators of public displays.

Regulations to underpin the new act will be brought in by November 2003 after a consultation period to be conducted by the DTI.

At present, it is not a legal requirement for firework or pyrotechnic operators to undertake any training although most reputable companies operate their own in-house training schemes and there is a BTEC qualification in existence that is a nationally recognised qualification. However, this is not yet compulsory and the course is not currently being run. Many people within the firework industry believe that a practical mandatory national training scheme is required which involves a 'logging' system documenting the exact level of competence for each operator. It is felt that this system is preferable to a classroom course as each type of firework material and all other conditions are logged, a bit like a pyrotechnician's CV. However, some theory training is also required. The scheme set up by the Guild of Firework and Pyrotechnic Operators combined with the BTEC Award fitted the bill perfectly and was recommended, yet various members of the GFPO could not see eye to eye and the scheme sadly fell apart. The new Fireworks Act may resurrect these ideas.

Now that The British Pyrotechnists Association and Firework Makers Guild have in place an Accredited Firer Scheme, I hope they do a little better.

To go into much further detail on fireworks and pyrotechnics is beyond the scope of this book but, needless to say, operators must be specially trained, risk assessments made and control systems put into place. Controls include safe firing and 'fall out' areas protected by stewards, fences, warning signs, marking and roping off, 'No Smoking' areas near fireworks, checks on weather, protective equipment and clothing for operators, etc.

Information about outdoor displays should be given to the local police, hospitals, ambulance, fire brigade, Air Traffic Control, Civil Aviation Authority, the Coast Guard, local institutions such as old people's homes and neighbouring landowners, especially those with animals and livestock.

With stage pyrotechnics it is vital that all crew, venue staff and performers are given instructions as to where the pyrotechnics are to be placed and when they are to be fired. The pyro must be kept in a locked room clearly marked 'Danger! No Smoking – No Naked Flame!' The key to the room must be held by the operator.

Some pyrotechnic devices containing ignition fuses require a police licence in addition to any public licence. This requires a named competent person and the correct lockable storage facilities for the pyrotechnics.

It is considered that electrical ignition is the safest method to fire pyro effects; the firing box should have an isolation or safety key that is kept with the operator. There is a theory that the emissions given off by mobile phones may be able to set off certain types of detonators and firework ignition devices. This is not proven as yet and more necessary research is underway.

PPE for firework/pyro operators includes: boots, safety helmet with visor, ear protection, flame/heat retardant gloves and overalls.

Stage Pyrotechnics are not covered by BS 7144; it is therefore possible for any person over 18 to purchase such devices. Each type of pyrotechnic does, however, require a document issued by the Health and Safety Executive known as the 'Competent Authority Document'. This and other C.O.S.H.H. information that enables an operator to carry out a risk assessment, should be available from a reputable manufacturer or supplier. If they can't or won't supply this information, change your supplier.

The main problems of pyrotechnics include prematurely triggering the pyrotechnic effect, use of larger quantities or more dangerous materials than needed, causing a fire, lack of adequate fire extinguishing capabilities, and, of course, inadequately trained and experienced pyrotechnics operators.

Training, a great deal of planning, and following the Code of Practice is essential to safe firework and pyrotechnic shows.

In February 2003 a fire broke out at a club in Minneapolis. Fortunately on this occasion no one was hurt. The cause of the fire was pyrotechnic devices being used by the support band; the club was not licensed for the use of pyrotechnics.

A week later a fire in a rock club in Rode Island USA killed 96 people and many more were seriously injured. Many of those killed or injured received their injuries whilst trying to escape and were overcome by smoke or were crushed. The cause of this fire was pyrotechnic devices used at the start of a performance by the band Great White.

It is understood that most of the one-story building was made from wood and the premises were again not licensed for the use of pyrotechnics. Fire-fighters concluded the whole building was consumed by flames in under than three minutes.

This last incident took place only a week after another serious incident at a venue in Chicago, where more then 20 people died while trying to escape from the venue in which stewards had used pepper spray and Mace in an attempt to control an incident. Those injured in this incident were crushed while trying to escape.

A Code of Practice for The Use of Pyrotechnics

This code of practice relates to the use of indoor or stage pyrotechnics and should only be used as a guide and not considered in any way definitive.

1. Pyrotechnics must only be used by competent adults. If misused, they can be very dangerous and cause serious injuries.
2. Always inform the venue/licensing authority of intent to use pyrotechnics at venues 14 days prior to the event. Include in the proposal a list of all devices that are to be used and the name(s) of the manufacturer(s).
3. The person responsible for the control and firing of pyrotechnics must ensure that it is safe to do so. The loading and sighting of all devices must be done with safety as the top priority. 'No smoking' signs should be displayed during loading of pyrotechnics and up to show time.
4. Firing systems used must be of a recognised design and manufacture, incorporating both electrical and mechanical failsafes, i.e. key switch and panel switch enable systems. The person in charge should carry the ignition key with him/her at all times and should only insert the key immediately prior to ignition. At no time should it be left in the controller.
5. Only use devices, hardware and controllers manufactured by a recognised and respected manufacturer.
6. It is illegal to mix/make up, charge or recharge devices on premises other than those licensed to do so. Therefore, all devices used must carry a Health and Safety Executive Competent Authority Document. The manufacturer should be in a position to supply copies.
7. All pyrotechnics must be stored in a safe and locked room, displaying 'No Smoking' until they are required. Only the devices required should be placed *in situ*; no other 'Spares' or such like should be present on the 'Stage'.

   All pyrotechnics must be delivered to the venue in packaging as defined by the Competent Authority Document.

8 When pyrotechnics are in use, the operator must have sight of all devices. If it is unsafe to detonate any device, he or she must not fire.
The final decision must rest with the operator. Suitable fire-fighting equipment should be deployed in strategic positions *in situ*, and stage technicians made fully aware of all devices, the position, and the proposed time of firing.
9 When positioning devices, always take into account all personnel, scenery and other objects, etc. *in situ*. Take notice and ensure that all fabrics in close proximity are of a non-flammable nature and that all staff, including stage crews and FOH stewards, are fully aware of the position and proposed firing times of pyrotechnics.
10 A suitable and sufficient risk assessment must always be carried out and the results recorded in writing prior to the use of pyrotechnics.
11 If any doubt exists about the suitability of any device, a test firing should take place with the relevant authority present so that a further risk assessment can be made. Many pyrotechnics' devices produce large quantities of smoke; the ventilation of the venue should also be considered as part of the risk assessment.

Lasers

We all know that water and electricity don't mix very well, so when using water-cooled lasers be sure the plumbing for the cooling system is of a good standard; large display lasers use a three phase electrical power, say no more.

The new safety standard for lasers is BS-EN: 60825 Part 1: 1994 *'Radiation Safety of Laser Products, Equipment Classification, Requirements and User's Guide'*. This replaces the old regulations and divides lasers into four main classes. Class 1 lasers are safe but all others can cause eye damage. In a nutshell, these regulations state:

1 The laser and any reflecting surface should be rigidly mounted.
2 The operator should be competent and be able to see the area of the effect.
3 A power cut off should be nearby.
4 All lasers other than Class 1 should be fixed so that their beams are 3m from the floor.
5 A competent laser safety officer, knowledgeable in the evaluation and control of laser hazards, should be appointed.

Crew should be instructed never to tamper with laser equipment and never to look directly into the beam of a laser; damage to the retina of the eye is permanent.

There is no direct legislation specifically for lasers, but present-day general Health and Safety regulations are relevant when using lasers. BS-EN: 60825 Part 1: 1994, *'Radiation Safety of Laser Products, Equipment Classification, Requirements and User's Guide'*, is the standard to be used by designers, manufacturers, suppliers and operators in order to comply with current relevant Health and Safety regulations.

Prior to public use, notification must be made to the local authority with information to demonstrate the system is safe and without risks to health. Details of the security arrangements and definitions of the conditions which would require the laser operation to be shut down, must also be available.

Purchasers of laser equipment should check the unit is EU compliant, has a removable key locking remote control system, proper safety warning signs/ stickers, detailed instructions and a safety shutter. When purchasing a laser make sure the supplier can show you (with a calibrated laser power meter) what the power output of the device really is, stickers and labels on boxes don't always mean what you will be getting!

Full details of notification details are contained in the HSE Guidance Booklet HS(G)95 *'Guidance on the Radiation Safety of Lasers used for Display Purposes.* This is an essential document for anyone involved in lasers and has replaced the HSE Guidance Note PM 19 *'Use of Lasers For Display Purposes'*.

With most of the special effects mentioned here, including lasers, fireworks and powerful spotlights, it will be necessary to inform local Air Traffic Control and the C.A.A. if this type of effect is to be used outdoors, especially near aerodromes.

Other Lighting Effects

A number of new high power luminaries can cause severe burns to the skin and eyes, rather like sunburn and 'arch eye' from welding. These effects can be prevented by never removing the cover and by over-riding the mechanical safety interlocks when the unit is switched on.

Changing bulbs on some luminaries can also be a dangerous process; the bulbs are high-pressure gas filled units that can explode with frightening results, in particular xenon and HMI bulbs. Only experienced operators who have been trained and are protected by full-face visors and long gauntlets, etc. should undertake such tasks.

The other point worth noting about these bulbs is that they have a limited life span and should be changed before their life is up or they may explode. It is for this reason that many intelligent lighting units have a meter on them to indicate the amount of use the bulb has received.

Xenon and HMI lamps of 5 kW and 7 kW are pressurised to about 8 bar when cold and around 30 bar when hot and so a lamp burst is possible with the resultant danger from flying glass. When the lamps are being installed, people should be made aware of the dangers and the vicinity of the projector or lamp housing cleared. This operation should not be carried out with the audience present. The lamp houses of commercial high-power projectors are designed to withstand a lamp burst, and to contain the glass within the housing even if the burst happens when hot.

Xenon and HMI lamps produce significant amounts of ultraviolet (UV) light. Commercial high-power projection systems are designed to contain this UV radiation within the projector and so should pose no risk to people if used properly.

The arcs of xenon and HMI lamps are very bright and housings are designed so that the arc cannot be viewed directly by the operator. Care should be taken that people are not put at risk by 'blinding' them with the light, especially if they are moving around in otherwise dark environments (e.g. while entering or leaving a venue).

General Requirements

All types of high power projection systems require significant amounts of electrical power, typically, 32 amps of three phase. Electrical systems that are installed for high-power projectors should take this into account and cabling should be rated accordingly. Sufficient dry powder or CO^2 fire extinguishers should be provided to ensure coverage of all the areas that house high power lamps and projectors.

Projection towers should conform to safe working conditions and careful enquiry should be made into the weight and dimensions of the chosen projection system to be used at the venue. Projectors being used externally need to be housed in a weatherproof projection cabin. Water must not enter the projector while in operation. Projectors need to be protected against unauthorised interference, and staffed or readily accessible by a competent technician at all times when in use at any venue.

34 LIFT TRUCKS, MEWPS AND PLANT

There are approximately 7000 reportable injuries (under RIDDOR) each year at work that involve lift-trucks.

The safe operation of lift-trucks and other plant depends on three key factors, all of which should be considered as part of a risk assessment:

- The driver or operator
- The truck or item of plant
- The area in which the operation takes place.

The Management of Health and Safety at Work Regulations call for operators of plant such as forklift trucks, cherry pickers, lift platforms and telescopic materials' handlers to be suitably trained and competent.

Evidence to prove the operator has received the required training, and is in fact competent, comes in the form of a certificate issued by an approved training organisation. However, companies may authorise their own staff and deem them competent to operate items of plant on there own premises after giving them suitable and sufficient 'in-house' training, for which records should be kept. Unless a nationally accredited training provider has provided training and certification, records must also be kept of the assessment. The records must also prove that the person who assesses the operator is also 'competent'. This written evidence must be submitted if and when required. Nationally accredited training providers are able to submit evidence of their trainer's qualifications if required.

It is not acceptable to send plant operators who do not hold a nationally recognised Plant Operators Training Certificate to work at sites or premises owned by other companies or persons, for example, when a staging company sends its staff to a festival site. In this instance the standards set by the promoter's (or whoever is in control of the site) Health and Safety Policy must be followed. These standards should call for a nationally recognised 'certificate of competence' to be provided.

There is an HSE Code of Practice and Supplementary Guidance for forklift operator training. This has been agreed following widespread consultation and is designed to help employers meet their obligations under the Health and Safety at Work Act in ensuring that all operators have the necessary training for Health and Safety. Within the code, the following organisations are recognised by the HSC as competent to accredit and monitor organisations to train forklift instructors and/or train, test and certificate forklift operators:

- Lantra (Sector Skills Council)
- Construction Industry Training Board (CITB)
- National Plant Operators Registration Scheme (NPORS)
- Road Traffic Industry Training Board (RTITB)
- Independent Training Standards Scheme and Register (ITSSAR)

The Joint Industry Council for Forklift Operating Standards and Vocational Qualifications has developed a standard skills test which has been adopted by all the above organisations. This test eliminates the confusion experienced in the past and means it's now possible to obtain an NVQ at Level 2 in Forklift and most other plant operations. The CPCS (Construction Plant Competence Scheme) card that is now also issued by Lantra, CITB and NPORS, requires the holder to fill in a log book and register a minimum number of hours before a card can be replaced. Under the CPCS, new cards are required every five years.

IPAF (The Industrial Powered Access Federation) offers specialist accredited training and certification to operators of Mobile Elevating Lift Platforms (MEWPS) and the organisations listed above can often offer training and certification for other items of plant and work equipment, such as angle grinders and abrasive wheels.

Holding a plant operator's certificate does not mean the holder is a 'safe' operator, but it does say the holder has received instruction and training and has been assessed and is therefore 'responsible' (and can be held responsible). An operator's certificate simply shows that the holder has received basic training - it is not a licence.

Operators need to maintain a level of competence to ensure they can use plant safely. Employers must have in place systems to monitor operators over a period of time after which they should be reassessed. These are the means by which the level of operator competence can be maintained.

Holding of a certificate to drive or operate one type of machine does not automatically qualify the holder of this certificate to drive or operate other types of machine. An operator with a certificate to drive an all terrain forklift truck cannot operate an industrial counterbalanced forklift truck or a telescopic materials handler. An operator must therefore be trained and assessed for each type of machine he/she intends to operate. Plant operators must be physically fit (in accordance with advice from the Employment Medical Advisory service) and not suffering from Epilepsy, Diabetes, Asthma, Heart or back disorders.

Users of powered access equipment and lift trucks are advised to be on their guard against illegal trainers and worthless certification. It is believed that a number of unscrupulous training centres are offering what they claim to be a nationally recognised certificate, knowing full well that, while the instructor is qualified to the required standard, the resulting certification is not. The HSE has reminded companies hiring powered access equipment and lift trucks that the Provision and Use of Work Equipment Regulations 1998 require the employer to ensure that equipment is operated only by people who are trained to operate it safely, and the only way that the standard of training can be guaranteed is when it is accredited by a nationally recognised training scheme.

Most lift trucks and their attachments supplied after 1st January 1989 must comply with the Self-propelled Industrial Trucks (EEC Requirements) Regulations 1988 which implement the EC Directive relating to self-propelled industrial trucks. The Lifting Operations and Lifting Equipment Regulations also cover lift trucks and MEWPS.

The use and maintenance of lift trucks and MEWPS are covered by the Provision and Use of Work Equipment Regulations 1998 which require all work equipment to be safe for its intended use, maintained in an efficient condition, and records kept to prove this is the case. From 5th December 2002, all mobile work equipment must comply with Part 3 of the Provision and Use of Work Equipment Regulations 1998, and in particular Regulations 26 (Roll-over protective systems (ROPS) and Regulation 27 (Overturning of Forklift trucks and restraining systems).

The manufacturer's or authorised supplier's instructions on inspection, maintenance and servicing must be followed, and records of inspection and servicing must be kept together with a daily 'operator inspection record sheet'.

If a machine or item of plant is hired then the hirer must be satisfied that the machine is safe before signing to accept delivery, in safe working condition. A copy of the operator's instruction booklet should be supplied with any hired machine or item of plant.

Under LOLER, lift trucks and mobile elevating lift platforms supplied on 'dry hire' must also have the records of test and inspection supplied to the hirer on delivery.

All lift trucks with a lift height exceeding 1.8 metres must be fitted with an overhead guard unless operating conditions prevent it.

To ensure the machine is safe upon delivery the following checks should be made. They are similar to the routine checks made on a car and should also be made by operators before each shift.

The inspection record must be recorded; a simple 'tick box' form is suitable for recording. Any faults must also be noted and reported, and any repairs effected before the machine is used, as well as conducting an overall visual check on the general condition of the machine or plant:

A plate containing certain information must be fitted to lift trucks and other

Horn	working and clearly audible
Lock	machine must be lockable with a removable key to keep it secure from unintentional use.
Brakes	in safe working order
Steering System	in safe working order
Tyres	correctly fitted, in good condition and (if pneumatic) correctly inflated. Correct tyre pressures should be clearly marked on the machine. Pneumatic tyres on rough terrain machines must be fitted with dust caps.
Lights (if fitted)	in good working order
All controls	should be clearly marked so that they can be seen from the operator's position.
Safe load indicator (if fitted)	must be in good working order.
Oil, water, battery, fuel	check for leaks and that all fluids are to correct level.
Hydraulics	visual check on hoses and connections for wear and leaks.

items of plant. This information includes:

- The name of manufacturer or authorised supplier
- The type of lift truck (or other plant)
- The serial number
- The unladen weight
- The capacity
- The load centre distance
- The maximum lift height

Employees have a duty to report all faults immediately, and employers have a duty to see that faults are repaired before that plant or machine goes back into service. Once again, records must be kept.

Lift trucks and are also covered by the Lifting Operations and Lifting Equipment Regulations 1998.

It is important to note that if pneumatic tyres are fitted to a machine they should be in good condition and inflated to the correct pressure. This is because the machine will not be stable with one or more soft tyres. The Lifting Operations and Lifting Equipment Regulations (LOLER) require that a method of checking the correct tyre pressure be provided.

Fork lifts and other items of plant must only be used for their intended purpose: they are not to be used for towing other vehicles out of the mud, etc. This would be inappropriate use of work equipment under the Provision and Use of Work Equipment Regulations 1998 and is an offence under the Health and Safety at Work Act.

Two common causes of accidents with forklifts are: a) the operator failing to look behind him to see what is happening and b) the operator being temporarily blinded by glare from the sun.

Working Platforms on Lift Trucks

Most forklift trucks are not designed to lift people. They may, however, be fitted occasionally with working platforms to provide a temporary place of work at height and so provide an enclosure to enable people and their equipment to be elevated.

People often need to temporarily work at height. Working at height is known as a high risk activity and the selection of a safe means of access is important. There is a variety of options available, such as the provision of a permanent staircase, erection of a scaffold or the use of a MEWP, tower scaffold or portable ladder. The selection of a safe method will depend on the work activity which is to be carried out, the frequency of occurrence, the duration of the work, and the availability of equipment.

Although forklift trucks are primarily designed for materials' handling, experience has shown that a forklift fitted with a suitably designed working platform can provide a safer alternative to other means of access, such as a ladder.

However, a conventional forklift fitted with a working platform will not provide the same level of safety as purpose-built equipment, such as a MEWP. It is therefore strongly recommended that, wherever reasonably practicable, preference be given to the use of purpose-built equipment.

The Law

As explained, the use of work equipment and how it should be used is covered by The Provision and Use of Work Equipment Regulations 1998. In addition, lift trucks are also covered by the Lifting Operations and Lifting Equipment Regulations 1998. The guidance for these regulations should be followed when using lift trucks fitted with working platforms.

A risk assessment should be carried out for the intended activities involving the use of a work platform with a forklift truck. The assessment should identify the measures needed to comply with the relevant legislation applicable to the activities carried out.

Working platforms are interchangeable equipment under the Supply of Machinery Safety Regulations 1992.

They need to meet the essential Health and Safety requirements of these regulations but are not an item of equipment covered by Schedule 4 of the Regulations so they do not need type examination by an approved body.

Basic Working Platforms – Design Features – Dimension

The dimensions of the platform should be as small as possible, compatible with the number of people which it is designed to carry and the work they are likely to undertake. The platform dimension parallel to the fork arms, i.e. forward length of the platform, should not exceed twice the rated load centre distance of the truck.

For front-loading trucks, the dimension at right angles to the fork arms, i.e. width of the platform, should not exceed the overall width of the truck by more than 500mm

Capacity of Working Platform

It is recommended that the platform be designed to carry no more than two people. To ensure correct use, the platform manufacturer should provide information on the maximum number of people and any limits on the additional loads which may be carried in the platform. This should be marked on the platform.

Floor

The floor of the platform should be of adequate strength, horizontal when attached to the truck in its elevated position, slip resistant, and designed to prevent the accumulation of liquid. Any openings in mesh floors should be sufficiently small to prevent tools falling through.

At any point, the floor should be capable of supporting a mass of 125kg applied over any area of 0.16 m^2 without permanent deformation and be capable of supporting without permanent deformation, a uniform pressure of 4500 N/m^2 over the whole area.

Rails and Toe Boards

All platform edges should be guarded by:

- A top rail: the upper surface being between 1000 mm and 1100 mm from the platform floor;
- A toe board: having a minimum height of 100 mm; and
- At least one intermediate rail equally spaced between the top of the toe board and the underside of the top rail.

Other equally effective means of guarding between the top rail and floor, such as infilling with wire mesh, panelling and/or safety glazing may be used.

To ensure that the rails have adequate strength, they need to be capable of withstanding, without permanent deformation, horizontal and vertical forces of 900N applied individually and concentrated at the points of least resistance.

Any infill would also need to be capable of withstanding, without permanent deformation, horizontal forces of 900 N concentrated at any point on its surface.

Gate

Any gate provided should open inwards, upwards or sideways and return automatically to the closed position. It is strongly recommended that the platform has a device which locks the gate automatically and ensures it cannot be opened once the platform is raised. If such a device is not fitted, then the gate should be self-locking in the closed position.

Handholds

Suitably sized and positioned handholds should be fitted within the confines of the working platform. To allow for safe use, e.g. to allow use with gloved hands, it is recommended that there is a gap of 90 mm between any handhold and the side of the platform.

Safety Harness Anchorage

Safety harness anchorages should be included on all working platforms. The anchorages should have sufficient strength to ensure their safety in use.

As a general guide, a single anchorage used by one person should be capable of withstanding 10 kN without any visible permanent deformation.

Protection from Moving Parts

Screens or guards should be fitted to the working platform to separate people being carried from any trapping, crushing or shearing points on the truck-lifting mechanism or any other dangerous parts, e.g. where a chain passes over a sprocket or between moving parts of the mast or its actuating mechanism.

To cater for people leaning on them, the screens and guards should be capable of withstanding, without permanent deformation, vertical and horizontal forces of 900 N applied individually and concentrated at any point. Greater strength would be required if the screens and guards were subjected to higher forces in use or if the separation distances referred to below are compromised when the 900 N forces are applied.

The screens and guards should be of sufficient dimensions to prevent people from reaching through, over and around them into hazardous parts on the truck.

The adequacy of the screens and guards may be assessed for particular trucks onto which the platform is intended to be fitted with reference to the separation (reach) distances given in British Standard BS EN 294:1992. Adequate information should be provided with the working platform to allow users to assess which trucks the working platform may be used with and to provide instructions on how to check that adequate separation distances are provided.

Securing a Platform to the Forklift Truck Fork Arms

Platforms for use on the fork arms of a truck should have fork pockets on their underside which will accommodate the fork arms spaced at the widest practicable distance apart.

A positive locking device, e.g. behind the heels of the fork arms, should be included on the platform to retain it on the truck when in use. Any loose components associated with the locking device should be secured to the platform so that they cannot be mislaid when the platform is not in use.

The fork pockets should be suitable for use with fork arms which vary in length from 75% to 100% of the platform forward length measured parallel to the fork arms.

They should fully enclose the fork arm along its length, i.e. they should not be open along their underside.

The working platform manufacturer/supplier should provide instructions to allow the platform to be fitted and properly secured.

Securing a Platform to the Forklift Truck Carriage

Information should be supplied with the platform to adequately identify the types of carriages to which it can be fitted.

Where a platform is designed to be attached directly to a truck carriage conforming to International Standard ISO 2328:1993, i.e. with the fork arms removed, the effacement points should conform to the requirements of 1S0 2328:1993.

If quick release bottom hangers are used, they be designed so they cannot become detached during operation and are kept near the platform to prevent them from becoming mislaid when it is not in use.

Locating Points

The locating (i.e. pickup) points on both carriage-mounted and fork arm mounted working platforms should be symmetrical about the centre line of the working platform.

Visibility

It is recommended that the platform should be painted in a conspicuous colour.

Warning Sign

A permanent notice should be fitted to the platform in a position where the truck operator can read it from his normal operating position, stating: 'before elevating the platform apply the parking brake and ensure the transmission is in neutral'.

Instructions should be provided with the working platform to inform truck operators to stay at the truck controls while the working platform is in a raised position

A warning sign should be provided in the platform to remind people not to leave the platform when it is elevated.

Identification Plate

A plate should be fitted to the platform clearly indicating the:

- CE Marking, i.e. declare compliance with the Supply of Machinery (Safety) Regulations 1992 (as amended).
- Name and address of the platform manufacturer or authorised representative.
- Serial number and year of manufacture.
- Un-laden weight of the platform and position of the centre of gravity.
- Lost load centre (if carriage mounted).
- Maximum allowable load in kg.
- Permitted number of people who may be carried.
- Minimum actual capacity of the truck on which it may be used.
- Truck and Working Platform Combination

The manufacturer of the working platform should supply information to allow appropriate trucks to be used with the working platform and to state the need for trained people to operate the truck.

Tilting Mechanism, Side Shift and Other Attachments

The platform manufacturer/supplier should supply information with each work platform to inform users that when a side shift or tilt mechanism is provided on the truck, to which the working platform is attached, the side shift should be locked in mid-position and the tilt mechanism locked so that the floor of the platform is horizontal when in use.

This can be achieved, for example, by fitting an appropriate locking device which disables the controls to the mechanism, or locking controls so they cannot inadvertently be operated.

Additional information should be provided stating that the working platform should not be used with trucks with other movable attachments unless they are locked. This may be achieved by a mechanical device or other means which in the event of failure, when the working platform is raised, fails to safety.

Selecting a Suitable Truck

General

It is essential that the work platform is compatible with the truck on which it is used under all operating conditions.

Before using any combination for the first time, the working platform and truck manufacturer's/supplier's information should be consulted to ensure that the truck and working platform are compatible.

If there is any doubt or if any modifications are required to the platform or truck, then the working platform and/or truck manufacturer needs to be consulted as appropriate.

Weight

The weight of the platform, together with its load of people, tools, materials etc. should not be more than half the actual capacity of the truck. The actual capacity of the trucks is the actual capacity for materials' handling at the rated load centre distance; in the case of telehandlers and reach trucks, it is the maximum lift height and maximum outreach. Account will need to be taken of any attachments and it should be remembered, that any items or parts placed on the platform while work is being carried out add to the weight of the platform. For VNA trucks with elevated operator position and secondary load elevating mechanism, the maximum lift height should be the maximum lift height of the truck with the auxiliary mast fully raised.

Method of Securing

The user should confirm that in accordance with the platform and truck manufacturer's recommendations, the platform is appropriate for use with the truck on which it is to be used.

The fork arms should preferably extend fully into the channels but should not extend less than 75% of the platform dimension parallel to the fork arms.

Guards

When the working platform is fitted on the forklift truck, the screens or guards on it should be checked to ensure that they provide adequate protection for people being carried. The screens or guards should provide adequate separation (reach) distances, as detailed in PS EN 294:1992 from danger zones such as the lift chains on masted trucks.

Inappropriate Trucks

Working platforms are not suitable for use on:

- Trucks with masts which can give erratic movement, e.g. due to sequencing problems during lowering.

- Trucks with an actual capacity (i.e. the capacity of the truck for materials' handling) of less than 1000 kg, unless their stability in use can be guaranteed under all working conditions, and rough terrain trucks with a lift height of more than 6m.

Note: working platforms may be used with rough terrain trucks at heights greater than 6m if they are of an integrated type, which together with the truck, meet safety requirements similar to a MEWP made to BS 7471:1989 or equivalent standards.

A Safe System of Work

Maintenance

Working platforms should be used and maintained in accordance with the instructions provided by the platform manufacturer.

Training

Truck operators and people expected to work on platforms should be properly trained. Full instructions on safe systems of work with platforms, including the action to be taken in the event of an emergency and the dangers associated with leaning out of the working platform, should be given.

Securing the Platform to the Truck

Pre-use checks should be carried out to ensure that the working platform is properly located and secured to the truck before each use.

Trucks

For all trucks covered by this guidance:

- The truck should not be moved while the working platform is elevated.
- The parking brake should be applied.
- Where applicable, the transmission placed in neutral before raising the platform.
- If the truck is rated for use with the working platform with its stabilisers deployed, then they must be deployed before the platform is elevated
- The truck operator should remain at the controls of the truck while the platform is in an elevated position.

  It is essential that the truck should only be used on firm, well-maintained and level surfaces. Gradients and uneven or inconsistent ground conditions can affect the stability of the truck.

People on the Working Platform

It is not appropriate for people to leave or enter the platform while it is elevated. In addition, it is advisable not to lean out of the platform when it is raised and, so far as is reasonably practicable, the platform should be positioned to prevent the need for people to lean out when carrying out their work. If people have to lean out of the platform then they should wear a suitable harness and lanyards linked to the platform's harness anchorage points.

There should be adequate communication between the truck operator and people on the platform. Handheld communication devices or an agreed system of signals can be used where communication is difficult, but account should be taken of any foreseeable hazards due to electromagnetic compatibility of, for example, trucks and radios.

If a working platform can be lifted to a height greater than 4m above the truck operator or the working platform is used in a noisy environment, then it is likely that communication aids, such as a whistle or klaxon, will be required.

Good communication is particularly important when the platform is raised and lowered. Extra care will need to be taken if more than one person is carried.

Some working platforms are fitted with controls, which, until held in position by the people on the working platform, prevent the truck lift/lower controls from functioning.

Such features on basic working platforms can add to safety in use and should be considered where communication between the people in the platform and the truck operator may be difficult.

Special precautions may be necessary to ensure that people in the platform are not endangered by hazards, such as live electrical conductors, etc.

For applications where overhead hazards exist (against which the platform or the people on it could be crushed), or where there is a risk from falling objects, a working platform with suitable overhead protection should be used wherever possible.

Tilting Mechanism, Side Shift and Variable Geometry Attachments

If the truck is fitted with a tilting mechanism, side shift or variable geometry attachments, then it should be ensured that the tilt, side shift or attachments cannot be moved while the working platform is raised. When using and installing the working platform on the truck, the working platform manufacturer's/ supplier's instructions referred to previously will need to be followed and account taken of the effect on stability of any attachments on the truck.

Segregation

Warning cones, light, or signs should be positioned around the truck working area where there is any possibility of other vehicles or pedestrians coming into close proximity, or of objects falling from the platform. Where necessary, other adjacent operations or activities may have to stop.

Verification of stability

It the dimensions of the working platform exceed those set above, the stability of the truck/working platform combination should be verified by testing or calculation, as appropriate.

Integrated Working Platforms

General

The safety of the basic type of working platform depends to a large degree on safe systems of work; safety can be improved if steps are taken to integrate the working platform with the forklift truck to which it is fitted. There is no standard means of connecting a working platform with the operating systems of a forklift truck, hence the provision of additional features, such as platform-mounted controls, will require modifications to be made to a designated truck, taking account of the truck manufacturer's advice.

The guidance given for basic working platforms on truck selection and safe systems of work applies as appropriate to integrated working platforms.

The restrictions on use already given, apply unless the working platform/ forklift truck combination gives an equivalent level of safety to a MEWP or if guidance already given is followed.

Controls

Control of mast tilt, carriage tilt, side shift, attachment rotation, truck travel and reach motion on masted reach trucks should not be available from the working platform.

Any controls provided should be located within the plan area of the working platform, returned to the neutral position when released, and be protected from inadvertent operation. Preferably, the controls should be positioned midway across the width of the platform and at the rear to keep the operator away from the edges of the platform while it is in motion.

Lifting and lowering should be possible from the work platform only by operation of a two-handed control. If the platform is to be used by more than one person at the same time then an operational set of two-handed controls should be provided for each person on the working platform. Lock out features may be provided on all but one set of controls to allow the working platform to be raised or lowered when only one person is using it, but should not be used for any other purpose.

When the controls in the working platform are activated, the controls in the truck should be deactivated, including travel motions. If the working platform is used with a truck which has stabilisers and the truck has been rated for use with the platform with its stabilisers deployed, then an interlock is required so that the working platform controls cannot be activated until the stabilisers are deployed.

An emergency stop control should be provided at or adjacent to the working platform controls, which will guarantee stopping of all powered motions when activated, even in the event of a hardware or control system failure.

Marking of Controls

The function of controls should be clearly and durably marked with graphic symbols in accordance with ISO 3287: 1978, and each symbol should be affixed on or close to the control to which it applies. The instructions should include a description of the function and location as fitted to the platform.

Over-riding Control Device

The truck should have an overriding control device to ensure that, in the event of an emergency, such as a power supply failure or an immobilised operator, the working platform can be lowered in a controlled manner to a safe position.

This control device should be clearly labelled with instructions for its use, and designed to be in a position where the person operating the control is not at risk from the descending platform.

Mobile Elevating Work Platforms and Safety Harnesses

Background

1. Mobile elevating work platforms (MEWPs) can provide safe access and

safe working at heights, and are often safer than ladders or other access equipment.

2. Each year there are a number of serious accidents in which operators are thrown from MEWPs, including about three fatalities each year from the baskets of cherry pickers. In many situations, the wearing of a safety harness would provide good protection in the event of falling, or of being thrown from the MEWP's basket.

3. This guidance relates primarily to cherry pickers. However, the advice is also relevant to scissor-type MEWPs if the operator is at risk of falling, e.g. as a result of leaning over the guardrail.

The Risks

4. The typical MEWP consists of one or more pivoted arms. Movement from a single pivot causes the basket to move through an arc. To make the basket move in an approximately straight line, i.e. down the face of a building, the operator must adjust more than one control, either simultaneously or alternately. Many accidents occur when the controls are too coarse, which may cause the basket to move more rapidly than anticipated or means that the operator may not be able to compensate sufficiently for the 'arc' movement. The basket may then stick or be obstructed by part of the structure, e.g. a stage roof. If power continues to be applied when the basket has become jammed, it is likely to cause failure of part of the supporting machine or it may resist in the sudden movement of the basket, throwing out the occupant(s).

5. Other significant causes of accidents are:

(1) When a nearby vehicle or mobile plant strikes the MEWP, e.g. if part of the supporting arm intrudes a public thoroughfare or contacts an overhead travelling crane.

(2) When the operator leans too far out of the basket or loses balance, e.g. when handling awkward pieces of material such as sections of truss.

(3) Failure of the levelling system or a major component of the MEWP.

(4) Unexpected movement or overturning due to incorrect installation, soft or uneven ground or incorrect tyre pressures.

6. Note that MEWPs are intended as work platforms and not as a means of access to elevated levels; persons have been injured climbing out of MEWP baskets to get onto a truss.

7. Error of judgment or lack of sufficient instruction and training and rapid movement of the basket and collision with racking, etc. While many incidents have been attributed to 'operator-error' many such errors are foreseeable and should be considered as part of a risk assessment.

Precautions to Be Taken Before Working from Any MEWP

8. Employers and others responsible for the use of MEWPs should assess risks of users falling from or being thrown from the basket, and take precautions to eliminate or control those risks.

9. The precautions for safe work from a MEWP include:

(1) Guardrails round the edge of the basket to stop the user falling.
(2) Toe-boards round the edge of the platform.
(3) Use of stability devices, e.g. outriggers provided to make the machine stable.
(4) Locking-out controls (other than those in the basket) to prevent inadvertent operation.
(5) A safe system of work, which includes:
  a. Planning the job (the Lifting Operations and Lifting Equipment Regulations 1998 [Regulation 8].
  b. Use of trained/experienced operator(s).
  c. Instructions when to enter/leave the basket, e.g. when basket is fully lowered.
  d. Instructions in emergency procedures, such as evacuation, should the power be lost.
  e. Use where necessary, of suitable fall restraint or in high-risk situations, fail arrest equipment.

High Risk Activities Requiring Fall Arrest Protection

10. In high-risk situations, the use of personal protective equipment (PPE) for fall arrest should be required as it could significantly reduce the risk of serious injury.

11. The equipment will usually consist of a full body safety harness attached by a lanyard (and energy absorber) to a designated anchorage point on the basket. Situations (including those described at paras 4 and 5) which have been identified as high risk are where:

(1) There are protruding features which could catch or trap the basket.

(2) Nearby vehicles or mobile plant could foreseeable collide with, or make accidental contact with, the MEWP.
Situations include work in the vicinity of very wet or slippery road surfaces, etc. where the MEWP may intrude the safety zone of a road traffic management section.

(3) The nature of the work being done from the basket may mean operators are more likely to lean out. This may happen, for example, when operators:
   a. Inadvertently or, for reasons of speed and convenience overreach or stretch from the basket and overbalance.
   b. Are handling awkward work pieces unexpectedly.

(4) Rapid movement of the machine is possible.

Appropriate Safety Equipment

Fall Arrest Equipment

12. Where fail arrest equipment is provided, it should consist of a full body harness (relevant standard: BS EN 361) and a lanyard. Inspectors may refer to BS EN 363 (fall arrest systems), which provides guidance on how the various components of a system should be assembled.

13. The safety harness should be attached to a suitable anchorage point in the basket (see BS EN 354) with energy absorber (see BS EN 355). An energy absorber may not be suitable for use with a MEWP which is restricted to low height working, e.g. below about 5 metres (depending on the fall arrest equipment used), as the distance between the anchorage point on the MEWP and the ground may not be sufficient for the energy absorber to deploy correctly before the user hits the ground. This clearance height should be considered in the risk assessment

14. The lanyard must be of correct length to allow normal work to be carried out without restriction, but should be as short as practicable. Operators will need training in the use of the harness and lanyard, and the procedures for periodic inspection, maintenance and storage of fall protection PPE (especially textile equipment).

15. Most MEWPs are fitted by the manufacturer with designated anchorage points (which should be separately marked). However, care will be needed when selecting a MEWP appropriate for the planned work, e.g. a retractable-

type fall arrester requires a high anchorage point. Some types of MEWP are not provided with an anchorage point and some have an anchorage port specified as 'fall restraint' only.

Fall Restraint Equipment

16. There will be a few scenarios, e.g. where part of the mechanism breaks, in which a safety harness attached to the basket will be ineffective as the workers may fall with the basket. This situation is, however, less common than one where the occupants are likely to be thrown out.

In many cases the basket is very unlikely to fall off, even though it may tip forward. A harness will still afford protection as long as the basket is higher off the ground than the length of the (extended) lanyard.

17. Where it is unlikely that a worker would be thrown or fall from the basket but the employer wishes to reduce the risk further, e.g. by deterring overreaching, fall restraint equipment may be used. This usually consists of a combination of a waist restraint belt (BS EN 358 work positioning systems) and a lanyard (BS EN 354).

There are as yet no applicable European standards for 'restraint systems', but the scope of BS EN 353 specifically states that such devices are not intended to arrest a fall.

18. Fall restraint equipment, e.g. a waist-belt, provides a lesser standard of protection and is not suitable for fall arrest; the restraint system must be such that it stops the worker falling in the first place. 'Restraint belts' are restricted to situations where, in conjunction with a lanyard, they prevent the wearer putting himself in a position where a fall is possible.

19. Inspectors may advise the use of a full body harness, lanyard and energy-absorber at all times, as the additional cost of a full body harness (compared with a waist-belt) or of an energy absorbing lanyard compared with a restrain lanyard, need not be excessive.

20. The employer's risk assessment should consider the degree of fine control necessary and available for the safe movement of the MEWP, the speed with which the basket could tip, and its likely final rest position, as well as the condition of the MEWP itself. Inspectors will review critically any risk assessment, covering the uses of a MEWP described in paragraph 10, which concludes that a harness is unnecessary.

21. Inspectors will draw the attention of relevant users and suppliers to these

requirements and take appropriate enforcement action where necessary. The benchmark for the risk of somebody being thrown from a MEWP is a remote risk of serious personal injury. Failure to provide a harness in the situations described in paragraph 9 is likely to result in a moderate risk gap.

Relevant legislation includes the Provision of Work Equipment Regulations 1998 (Regulations. 4, 5,9, 15 and 20, and the Construction (Heath, Safety and Welfare) Regulations 1996 (Regulation 6).

BS EN 361:1993 Personal protective equipment against falls from a height. Full body harnesses.

BS EN 363:1993 Personal protective equipment against falls from a height. Fall arrest systems.

BS EN 354:1993 Personal protective equipment against falls from a height. Lanyards.

BS EN 355:1993 Personal protective equipment against falls from a height. Energy absorbers.

BS EN 358:1993 Personal protective equipment against fails from a height. Work positioning systems.

35 TRAFFIC MANAGEMENT AND WORK PLACE TRANSPORT SAFETY

The transport sector is one of the most regulated areas of any industry. Not only are there tacograph regulations for drivers, but the Working Time Regulations also affect them. Holders of Operators' Licences for goods vehicles and/or transport companies are required to ensure that drivers have sufficient statutory rest breaks during journeys; drivers are legally required to ensure they take their statutory rest breaks.

Every site operating a road transport business is required to have at least one person (usually a manager) to act as Transport Manager.

The essential qualification for a road transport manager is the Certificate of Professional Competence (CPC) of the Department for Transport, Local Government and Regions (DTLR) – Department of the Environment in Northern Ireland – which equates approximately to NVQ level three. There are four types of certificate: National or International Road Haulage Operations and National or International Passenger Transport Operations.

There is no other specific qualification requirement to enter work as a road transport manager.

Road transport managers have responsibility for planning the routes and schedules of drivers involved in road haulage, distribution and logistics or passenger transport. They must ensure that all operations are carried out in accordance with UK and EU laws and regulations governing vehicle safety, environmental controls on fuel emissions and traffic congestion, driver hours, customs requirements, and food hygiene, where applicable.

I will not dwell too long on the Road Traffic Act, but there are regulations not just for drivers of vehicles but for those who have to work on the highway, such as anyone involved in directing traffic, erecting signs or possibly loading/ unloading vehicles. Some of these regulations fall under The New Roads and Street Works Act 1991. Traffic management staff should acquaint themselves with Chapter 8 of the Road Traffic Act. All Traffic Management signs (both on and off site) should meet the requirements of the Traffic Signs Regulations (as shown in the Highway Code), but it must be remembered that traffic direction on public highways can only be carried out by police officers and traffic wardens: it is dangerous and illegal for anyone else to direct traffic on public roads. Persons directing traffic not on public highways must be aged over 18 and adequately trained.

High visibility jackets of the required standard are required for anyone working on the public highway or in car parks. They may also need PPE to cope with exhaust fumes and noise caused by passing traffic, and anyone directing traffic will need to be suitably trained and competent.

Any work on the public highway will almost certainly come under the New Roads and Street Works Act 1991; this includes the erection of signs. To comply with this act will require trained and qualified personnel.

Event organisers are advised that directional signs will probably need planning permission. Motoring organisations such as the RAC or the AA are the best people to approach for arranging directional signs to sites and events. These organisations are able to arrange the necessary permissions and have the required training and qualifications to carry out such an operation. You can be sure the signs these organisations erect will comply with the all the relevant Safety Signs Regulations.

The movement of trucks and vehicles (including motorcycles) on the public highway, an outdoor or festival site or even in the yard of a hire company, is a risk that is significant enough to warrant a Risk Assessment. At least one fatality has occurred due to poor management of vehicles on a festival site in recent years and hundreds of accidents have taken place in company yards and warehouse car parks, but the highest level of work related driving fatalities are on the public highway.

Like all risk assessments we must first consider the range of potential hazards; this will be wide-ranging. Talking to drivers will bring first hand experience of what happens in reality, but the views of those who only use the roadways occasionally should also be sought, since different views and answers may be given.

The risk assessment for at-work related driving hazards should consider passengers, pedestrians and other road-users, as well as the driver of the vehicle. Young or inexperienced drivers are particularly at risk, as are those who drive long distances.

Evaluating the risk is the next stage. You will need to consider how likely it is that each hazard will cause harm.

This will determine whether or not you can remove the hazard or if more action is required to reduce the risk. Some risks will remain after all precautions have been taken: decide if these remaining risks are acceptable.

Can the Hazard be Eliminated?

Is it really necessary to make the journey? Is it possible to hold a telephone, computer or video conference? If the hazard can't be removed, then controls must be introduced to reduce the likelihood of a road traffic accident involving an employee.

Control Systems

Can the journey made in full or in part by train?

Are schedules and journey distances realistic? Drivers must not be put at risk due to fatigue as a result of excessive driving distances. Drivers must be given adequate time to complete journeys safely; this is particularly important when scheduling tours.

Are company and fleet vehicles chosen for their safety features such as airbags, side airbags, impact protection systems, head restraints, anti-roll bars, lights, horn, reflectors, seatbelts, ABS breaking, etc?

These features need to be considered and assessed, because vehicles are covered under the Provision and Use of Work Equipment Regulations 1998 (PUWER '98):

- Are all vehicles properly serviced and maintained?
- Do you have adequate service and maintenance records?
- Are seatbelts and airbags fitted, tested and fully serviceable?
- Do staff use them?
- Are helmets and protective clothing provided for motorcyclists?

A number of other factors should be taken into consideration:

- Drivers' competence – are they aware of company policy on work related road safety and do they understand what is expected of them?
- Are all drivers suitably trained and do they have adequate experience?
- Would you expect a non-vocational driver to drive and work longer than a professional driver?
- Do they know how to carry out routine safety checks and adjustments to their vehicles?
- Do the drivers of high-sided vehicles know the height of their vehicles both laden and empty? There are around three to six major bridge strikes every day!

- (Vehicle height has to be displayed in the cab and the un-laden weight of the vehicle has to be displayed on the vehicle. It is the driver's responsibility to ensure he does not travel on the public highway over the gross legal weight).
- Do drivers know what actions to take to ensure their own safety following a breakdown of their vehicle?
- Do drivers know how to use anti-lock brakes (ABS) properly?
- Are drivers fit to drive?
- Are they taking medication that may impair judgment?
- Have they adequate sleep and rest?
- Do they suffer from stress? Stress can lead to road rage.
- Do drivers know what to do if they feel sleepy?
- Do they know the dangers of fatigue?
- Have they suitable and adequate nourishment?
- Are drivers provided with safety information about the vehicle, such as tyre pressures?
- Have you specified what standards and expertise are required for the circumstances of the particular job? How do you know these standards are met?

It is rare for these factors to be taken into consideration or implemented by a company; any accidents could be prevented if they were.

Routes should be thoroughly planned and suitable for the type of vehicle being used. Supplying drivers with the relevant maps helps; a photocopy of a route plan is not really suitable as diversions often occur and the driver can be forced down an unknown route without a map.

Weather conditions also need adequate consideration when planning journeys. Conditions such as snow, ice, rain, mud and high winds may require extra safety devices to be provided and additional training for drivers; a radio with regular traffic bulletins can be invaluable.

The assessment should be recorded in writing, monitored and reviewed as necessary.

The controls that can be introduced in car parks, yards and outdoor sites include speed restrictions, signs, road markings, lighting, well-marked traffic lanes and parking areas.

In order to provide a 'safe place of work', event organisers and production managers must control vehicle movement on site. Vehicle movement must be restricted to essential operations only once the public has been admitted.

Furthermore, a system must be in place so that all vehicles must report to gate staff before entering the site, and gate staff should give specific directions and information that drivers must follow while on site. Suitable signage, fencing and traffic stewards all play an important role in safe traffic and vehicle management.

Stewards involved in traffic management and directing traffic should not only be trained in a hand signal system that drivers will be familiar with, but they should also be trained in the manoeuvring characteristics of the vehicles to be directed.

On 'green field' sites, provision must be made for sufficient hard standing and turning places for vehicles in the event of wet weather, and adequate access controls (pass systems) must be in place together with information for drivers such as signs, maps and stewards to control and monitor vehicle movement, as we have already mentioned.

I make no apology for mentioning again that it must be a condition of entry that all vehicles report to the gate/site/production office for instructions before entry to the site. Once on site, it is vital that everything possible is done to separate pedestrians from vehicles by the use of railings, fences, or other suitable means.

Some indoor venues allow vehicle access. In these circumstances, always turn off the vehicle engine as soon as possible and check the building is well ventilated to prevent the build up of dangerous exhaust fumes.

Employers should keep records of employees' driving licences (especially if they drive company vehicles) together with insurance details and vehicle servicing and maintenance records as evidence and proof of competence. Employers should make regular checks on staff driving licences to check they are still valid and cover the relevant vehicle types.

Of course we know that we don't need a driving licence to drive a vehicle for fun or pleasure off the public highway on, say, a showground or festival site, but if you are driving in these places for work, then the HASAWA applies and this requires workers to be trained, competent and qualified (where a qualification exists), and the qualification and proof of competence to drive a vehicle is, of course, a driving licence!

Not only does a driver need a driving licence to drive a vehicle on a site, but the vehicle itself must be in a safe condition (PUWER '98), so you may need an MOT certificate to prove this! The days of daggy old hippy vehicles with the roof ripped off for use as a site vehicle over are well over!

Anyone who drives a private vehicle (or is employed to drive a vehicle) in the course of their work or business must be certain that they are covered by motoring insurance to drive a vehicle for all work or business purposes, and not just covered for social, domestic and pleasure use! 'Runners' for 'local crew' companies and venues be warned!

Unfortunately, motor insurance companies usually impose a heavy 'loading' on motor insurance for people who work in our industry.

Further problems with insurance can often arise with the use of self-drive vehicles and in particular, splitter buses. Tour managers and crew who 'double up' as drivers should check they are fully covered by their own or the hire company's insurance. In fact, anyone who uses transport services should check that the proposed freight/trucking/bussing companies hold all the required insurance for goods in storage/transit and driver/vehicle insurance; this must be valid for all the territories to be visited.

All too often I have seen good, careful drivers drive at 100 mph through pedestrians when they are on festival sites. If a flashing beacon is fitted to a vehicle if should be switched on when driving on an outdoor site, or the headlights may be used. On no account must hazard warning lights be used on a moving vehicle as it makes it impossible to tell if the vehicle is indicating to turn, and this has been the cause of several accidents.

A speed limit of a maximum of 10mph should be set on all outdoor sites and special care must be taken with reversing vehicles, especially in loading docks.

Some loading docks have an escape refuge where a worker can escape being crushed by a reversing vehicle; these refuges must be kept clear of rubbish and items such as oil drums.

Under no circumstances must people be allowed to stand in or on moving vehicles and passengers must not be carried on vehicles (such as fork-lifts) unless the vehicle is designed to carry passengers.

When loading or unloading trucks, be sure that the vehicle is on firm level ground and that there is sufficient lighting.

Truck drivers are generally responsible for their vehicle 'pack', so if drivers get involved in the loading/unloading of trucks, they should have received correct instruction in manual handling and be equipped with the required PPE, such as safety footwear.

When the doors of a truck are to be opened, all crew should stand well clear and the truck driver should open the doors while remaining behind the opening door. The reason for this is that if the load has moved during transit, the driver is the only person who is likely to be aware of this. Crew must stand well clear when the doors are opened so that loose equipment does not fall on them. Truck drivers should remove keys from the ignition beforehand and give them to a senior supervisor (who will be present during the whole loading/unloading operation) before opening the truck doors. Vehicles should not be moved with a part load unless it is secured or, in the case of a van, the doors are closed.

On festival and outdoor sites it is quite common for the fire officer from the local authority to set an Entertainment Licence condition that states that vehicles must not be parked on roadways or tracks, as they are often access routes for emergency vehicles or escape routes.

For fire prevention, it is also common to find that vehicles are not allowed near tents, campsites and temporary structures, and even awnings can be banned from vehicles on some sites. The reason for the ban is that if the nylon used in many tents is put together with petrol and oil, it can produce Napalm. Campfires are usually, and should be, banned in car parks. Drivers must be instructed to park in the areas designated for them even though this may cause some inconvenience to them.

As a condition of an Entertainment Licence, the emergency services will often insist on one route on and off an outdoor site being designated for emergency vehicles only.

Fork lifts and similar items of plant must never be used to tow other vehicles through muddy fields or for any other form of towing. Until recently, a crew person was awaiting trial for possible manslaughter of another person working on the same site, after a fork lift, being driven by the crew member, ran backwards and crushed the other person against a lorry when they were attempting to attach tow ropes to the lorry and tow it out of the mud with the fork lift. The case never made it to court on a technical point.

This prosecution was for a breach of the Provision and Use of Work Equipment Regulations. The regulation states that work equipment must be suitable for its intended purpose; a fork lift is not designed or considered suitable for towing other vehicles.

Many trucks used in our industry are fitted with tail lifts and are covered by the Lifting Operations and Lifting Equipment Regulations 1998 (LOLER). However, the three point linkages of an implement lift fitted to the back of farm tractors are not covered by LOLER.

Tractors and other vehicles with forklift attachments; cranes and HIABs are also covered by LOLER. Forklift attachments, cranes or HIABs must be marked with their safe working load (SWL) or Working Load Limit (WLL) and the records of test and examination must be available to the hirer/operator who will be classed as the 'duty holder' in a 'dry hire' situation under LOLER.

For those who own their vehicles with these attachments, they are reminded that the attachment must be marked with its safe working load and that tests and thorough examinations are required by LOLER at least every 12 months, and that detailed records must be kept. (See LOLER)

The use of All Terrain Vehicles, such as quad bikes and tricycles, is now becoming quite common amongst staff on large outdoor sites. Some of these machines are very powerful and require extreme care in their use. Many recent accidents involving these vehicles have been due to overturning on slopes and are making the HSE consider compulsory fitting of roll bars or cages to these machines.

At the present time the use of roll bars or cages is still under review and the HSE's only available means to tackle the problem is by advice, information and enforcement aimed at suitable training and the wearing of motor cycle crash helmets (to BS 6658 and BS 4110) in particular.

Goods Vehicle Operators' Licences

To find out if you require an 'O' Licence (Operators' Licence), try answering these questions:

1. Do you use a vehicle of more than 3.5 tonnes gross-plated weight?
2. Do you use a vehicle that has no gross-plated weight, but an unladen weight of more than 1525kg?

If you have answered "yes" to any of these questions, then you probably need an Operators' Licence, as do most of us!

Operators' Licences are available in three different formats:

- Restricted
- Standard National
- Standard International

The 'Standard National' licence will also require that the holder of the licence or a member of the company holds a CPC (Certificate of Professional Competence).

With the 'Standard International' licence, the holder or a member of the company must hold an International CPC.

To carry your own goods in the course of your normal business in Great Britain you need a 'restricted licence'. This does not allow you to carry goods for other people for hire or reward. *If you do, you could lose your licence and face a fine.*

A 'standard national licence' allows you to carry goods in Great Britain as well as the goods of other people for hire and reward in Great Britain. You will still need the 'standard national licence' to carry other people's goods for reward, even if it's only the odd day that you do this. This licence does not allow you to use your vehicles for international hire or reward work.

A 'standard international licence' allows you to carry your own goods and goods for hire and reward both in Great Britain and on international journeys.

You can obtain the full information on all the above licences and how to obtain the licences from the Department of the Environment and the Regions as well as from your local traffic office.

The laws on hiring 7.5 tonne trucks have also changed. It is now an offence for a hire company to hire a 7.5 tonne to a company that does not have an operators' licence.

If it is an individual wanting to hire the vehicle then he has to sign a form stating that the vehicle is not being used for reward. Should the individual be found using the vehicle for reward, they would face a fine of about £2500.

Once you have your 'O Licence' it's worth remembering that you will have to stick to the rules laid down by current tachograph legislation, as well as maintenance schedules if you happen to own the vehicle.

Drivers must not be put at risk due to fatigue as a result of excessive driving distances.

Drivers must be given adequate time to complete journeys safely, this is particularly important when scheduling tours.

One road traffic accident could cost a whole tour, not to mention peoples lives!

36 TEMPORARY STRUCTURES

Outdoor stages, mixer towers, PA wings, grandstand seating, viewing platforms, marquees, tented structures, dance platforms, pit barriers, platforms and masts for lighting, sound and cameras as well as structures to support video screens, are all classed as temporary demountable structures.

Full details on the design, procurement and use of temporary structures can be found in the Event Safety Guide and the Second Edition of the Institution of Structural Engineers publication, Temporary Demountable Structures. These two publications are considered essential reading for anyone hiring, building or using temporary structures.

Promoters, production managers, and others using temporary structures must allow sufficient time in production schedules for structures to be erected safely. The weather and ground conditions should also be taken into consideration as they can have serious implications on schedules.

When contracting a supplier of temporary structures, the usual checks should be made to ensure the competence of the contractor. Additionally, the contractor should supply design drawings, structural calculations, statements of loading, and method statements, as applicable for the proposed structure. Fire safety certificates may also be required for any fabrics used on the structure. The contractor supplying a temporary structure should provide a 'Completion Certificate' to confirm that the appropriate independent erection checks have been carried out and that the temporary structure has been erected in accordance with the design drawings and documentation. Where the erection checks carried out by a member of the erection team, evidence of that person's competence should be made available. Clients and Licensing Authorities may wish to have independent design checks carried out and these should be by chartered structural engineers with experience and skill of such structures. The Performance Textiles Association (formally known as MUTA, the Made Up Textiles Association) Code of Practice should be followed as a framework for design checks on marquees and tents.

I have seen a number of risk assessments dealing with the erection and dismantling of temporary structures that have not been 'suitable and sufficient'. These assessments often avoid some of the real problem areas; these problems usually involve safety with working at height and lifting operations. It has been noted that a number of staging companies have been cutting costs by not providing the ballast or kentledge required to assist with the stabilisation of the structure and to prevent overturning. The advice is to always carry out independent checks to see if kentledge or ballast is required, particularly if the supplier insists that it is not required yet cannot support the claim.

The supplier of a temporary structure should have 'competent persons' remain on site with a temporary structure to carry out supervision, in-service inspections and maintenance of the structure during the event.

The Institution of Structural Engineers coordinates the Advisory Group on Temporary Structures (AGOTS) whose objective is to persuade local authorities to use the best documentation available when assessing events with a view to reaching informed dialogue between inspectors and event staff. This group is also working on a new standard for Wheelchair User Platforms and from time to time offer formal advice on various aspects of design, procurement and use of the many varied types of temporary structures.

Those supplying temporary structures will almost certainly be subject to the Work at Height Regulations, LOLER, PUWER and the Manual Handling Regulations, to name but a few.

Finally, the HSE has advised its inspectors of the unsuitability of Genie Towers (Super Lift Towers) to suspend speakers at outdoor events following a number of accidents. Genie Towers are designed to be used indoors and in pairs to support trussing, etc.; they are not designed to support speakers outdoors.

A temporary structure may be very safe after it has been erected, but the methods of erection and dismantling employed by some suppliers are often very unsafe and expose those engaged in the erection and dismantling operation to serious hazards and high risks.

These hazards must be identified, assessed and 'suitable and adequate' control systems implemented to meet minimum statutory requirements. This will reduce ricks to an acceptable level.

37 FIRE SAFETY REGULATIONS

A guide to the minimum requirements for fire-fighting equipment and means of escape for permanent and temporary venues covered by an Entertainment Licence is set out in the Home Office publication, 'A Guide to Fire Precautions in Existing Places of Entertainment and Like Places' (also known as the 'Yellow Guide' or the 'Primrose Guide'), and in the Event Safety Guide, also known as the 'Pop Code' or the 'Purple Guide'.

The conditions attached to an Entertainment Licence may require additional fire precautions to be taken. For a permanent venue it is usually the occupier of the premises who holds the Licence that must be renewed annually. For a Temporary Licence, is it normally the Promoter (who may or may not also be the occupier) who holds the Licence.

It is usually a condition of Entertainment Licences that copies of Test Certificates for flame proofing and/or flame retardants for any fabrics used in the production or on site (tents, marquees, temporary stage coverings, drapes, etc.) are made available for inspection. Fabrics meeting the standards set by the Performance Textile Association (formally MUTA, the Made Up Textile Association) and carrying the "MUTA" or the new "PERTEXA" Approved label are generally very acceptable to Fire Officers and Inspectors. It is also a common Licence condition that all paint used on scenery and flats is water-based (emulsion) and hence less flammable than oil-based gloss paint. Petrol is also a banned item because it has a much lower flash point than diesel fuel and is potentially very dangerous.

The Guide to Fire Precautions in Existing Places of Entertainment and Like Places is an excellent source of reference material and is essential when planning events and carrying out fire risk assessments. While a little dated, it is used by fire officers when considering what appropriate conditions and restrictions to apply to an Entertainment Licence.

The requirements for our offices, warehouses and factories are detailed in The Fire Precautions (Workplace) Regulations 1997 and The Fire Precautions (Workplace) (Amendment) Regulations 1999. *You are required under these regulations to make a Risk Assessment of your workplace with regard to fire safety, numbers of people at risk, flammable or combustible substances, means of escape, means of raising the alarm and of your fire-fighting equipment requirements.*

In short, under the Fire Precautions Act 1971, any office, shop, factory or warehouse where 20 or more people are employed, or where ten or more are employed elsewhere but on the ground floor, requires a Fire Certificate. Some premises where fewer than 20 people are employed will also require a Fire Certificate in multi-occupied premises. If certain hazardous or high-risk operations are carried out, such as in a firework factory, or if means of escape may be difficult, such as in a multi-story building, then an application for a Fire Certificate should be made to the County Fire Brigade who will forward standard form (FRD 1) for completion. On receipt of the form they will arrange for an inspection to take place. *Please note that once the application has been made and acknowledged by the Fire Brigade you are deemed to comply with the Act.*

It is beyond the scope of this title to go into detail about the requirements for Fire Certificates but information may be obtained from some of the publications listed for further reading in the back of this book.

By 2005 it is planned to repeal all existing fire safety regulations and produce one set of regulations to replace the existing ones.

In common with other changes in Health and Safety legislation, this will be a result of European Directives as well as common sense. This is a very welcome move and will do much to simplify the whole current situation.

To obtain insurance on a workplace will normally require you to comply with the conditions for fire safety set by whatever insurance company you are insured with. The insurance company will probably send an Inspector round each year to visit, therefore there is little need for me to go into detail about what is required in terms of equipment and means of escape. In addition, your local Fire Brigade can provide information on fire safety and the Fire Precautions Act 1971.

The Fire Precautions (Workplace) Regulations 1997 and The Fire Precautions (Workplace) (Amendment) Regulations 1999 require employers to:

- Assess the fire risks in the workplace (either as part of your general review of Health and Safety which you already carry out or, if you wish, as a separate exercise).
- Check that fire can be detected in a reasonable time and that people can be warned.
- Check that people who may be in the building can get out safely.

- Provide reasonable fire-fighting equipment.
- Check that those in the building know what to do if there is a fire.
- Check that you maintain your fire safety equipment.

Informing staff of the procedures and systems in place in the event of a fire, including the means of escape and method of raising the alarm, should form part of your staff Health and Safety induction training.

Providing this information is also a requirement under Regulation 7 (Procedures for serious and imminent danger and for danger areas) of the Management of Health and Safety at Work Regulations 1999.

Early in 1999 the Fire Precautions (Workplace) Regulations 1997 were amended to become the Fire Precautions (Workplace) (Amendment) Regulations 1999. The main reason for this was to remove the overlap on enforcement between these regulations and the Management of Health and Safety at Work Regulations 1999 and thus put them in line with European standards (Euro Directive). It also confirmed the unconditional nature of an employers responsibilities and to remove the exemptions that existed in the previous regulations.

Everyone must be aware and staff must be informed that fire-fighting equipment, alarms, signs and fire escape routes must be uncovered and kcpt clear of obstruction at all times and that fire doors must be kept shut but not locked. They should also be aware that if any fire-fighting equipment is used, then its use must be reported as soon as possible so that it can be replaced/refilled.

Upon arrival at a new venue it is good practice to familiarise oneself with the position of all means of escape, fire-fighting equipment, alarms and procedures. A good production manager/crew boss should ask that you do this as a matter of course.

It is an employer's legal responsibility to instruct all staff and be certain they are familiar with evacuation procedures, fire safety and very basic fire-fighting techniques. This is yet further information that must be recorded in case it is needed as evidence of staff training and competence.

Indoor venues should display a plan of the venue just inside the doors; this is of great use to visitors, staff and fire fighters in case of an emergency.

Finally, production managers/venue management should insist that everyone checks into or out of the venue. This is not a clocking-in or out system for wages, it is simply a safety procedure because we don't want to send fire fighters into a blazing venue to look for you if you're just down the pub on a lunch break (not that you should be drinking alcohol when working of course).

Employers and venue owners who have a responsibility to provide fire detection and fire-fighting equipment should seriously consider a maintenance contract from an approved supplier who will check, maintain and test all equipment and to test certificates for extinguishers on a regular basis. As well as easing the burden on one aspect of your record keeping, this will keep you within the law. Indeed it may even be a condition set by your insurance company that this is in place before insurance cover is provided.

A recent change brought about in order for us to fit in with the rest of the European standards, was to make all fire extinguishers red, regardless of content. It is my belief and, indeed, the belief of many fire officers from various fire brigades I have spoken to, that this was an unnecessary and perhaps unwise change to the regulations.

In the past we all knew (or should have known) that a red fire extinguisher was for water, a black one for carbon dioxide, a cream one was foam, blue was for dry powder, etc. Now all new extinguishers must be red.

We now have to spend time reading the label to see what the extinguisher contains so we know what type of fire to use it on. In the meantime, the building burns down around us! As said, even Fire Safety Officers from the Fire Brigade are unimpressed by these new regulations.

Well, it may not be so bad – those clever European bureaucrats in Brussels have said we can have a little black, cream, blue or green 'flash' on the extinguisher to help us. Thank you chaps, but it's still going to take time to check and that time may cost lives.

As I am sure you all know, fire needs three elements to start: fuel, oxygen and heat (the Fire Triangle); remove any one of these elements and the fire goes out. Oxygen is all around us in the air we breathe and fuel can be just about anything to a fire. We can stave the fire or stifle it of oxygen by covering it with a fire blanket or using a carbon dioxide extinguisher, we can cool the heat with water and we can stave it of fuel, but never try to move a burning fire to remove fuel. That is asking for trouble.

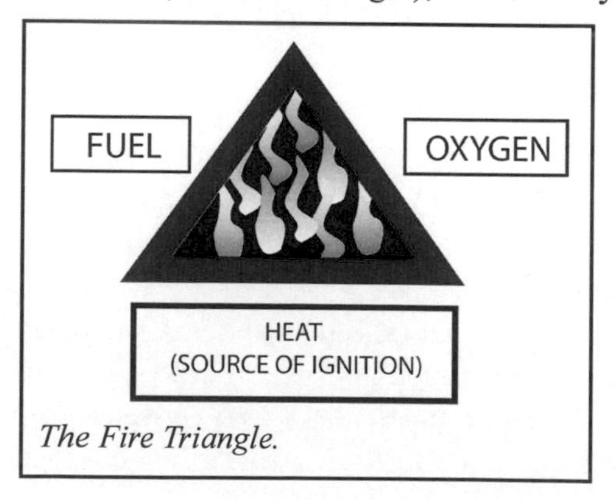

The Fire Triangle.

Causes of Fire

There are only two causes of fire: people and nature. All fires can be attributed to one of these two causes.

Fire Spread

The horrific fire some years ago that destroyed the stand at Valley Parade, home of the Bradford Football Club, did the majority of its damage in less than six minutes.

Whilst this fire (that killed approximately fifty-eight people) was in a football stand, it could quite as easily have been at an outdoor stadium-type rock concert or any other large public gathering. The cause of this fire was attributed to build-up of litter underneath the stand that was then ignited, quite possibly by a burning cigarette end. This tragedy also highlights the need for monitoring as part of the risk assessment process. Many of the fans at Valley Parade were unable to escape because most the emergency exists had not been unlocked prior to the fans' admission to the ground. Furthermore, the fire escapes were locked to prevent illegal ingress of non paying public. The risk assessment showed that it was essential to unlock the emergency fire exits but this was not done because it involved an extra cost to have stewards man the unlocked gates and so it was decided to keep them locked until near the end of the game. This highlights the need to have effective monitoring in place by persons who understand safety management systems.

That incident took place a few years ago and since then a lot of work has gone into training security staff in fire safety and evacuation techniques as well as introducing risk assessment and monitoring systems. Good stewards are essential to a well-run event and production crews could help a lot more by showing security staff more respect and treating them as fellow, trained professionals.

Fire Prevention

Good 'housekeeping' is paramount to fire prevention. Keeping your workplace clean and tidy, preventing unauthorised access, observance of 'No Smoking' signs, correct use of electrical equipment, safe use of chemicals, and only using fabrics and materials that have been treated to make them flame-retarding or which are inherently flame-retardant. This includes lighting gels and all drapes which should have a Fire Safety certificate. All of these measures provide a basic means of fire prevention.

Part of our 'housekeeping' is the prevention of litter and waste. Not only is litter and waste a place for germs and disease to propagate it's an attraction for vermin such as rats that spread diseases such as Leptospiriosis. For this reason:

- Always keep your workplace clean and tidy and use gloves to handle waste.
- Clean up spillages promptly, especially chemical spillages.
- Never dispose of waste in streams or watercourses.
- Never pour chemicals down the sink or dispose of them on a fire, especially aerosol containers!

We have already mentioned the need to keep access roads clear around festival sites, as they may need to be used as emergency roads, as well as the requirement not to have vehicles in campsites to prevent the spread of fire.

When the nylon used in the construction of lightweight tents mixes with petrol or oil, it can form the flammable chemical Napalm; so keep cars and tents separate! The only safe fuel allowed on most outdoor sites is diesel, and even then it must be stored correctly.

Means of Escape and Raising the Alarm

The Fire Brigade must be called to every fire no matter how small and even if you think the fire is out.

Fire doors are important components for containing fires and likewise many theatres have an Iron (Safety) Curtain to do the same sort of job. It's crucial for Fire Doors, Safety Curtains and all fire-fighting equipment to be kept clear of any kind of obstruction.

Do not use the lift (if the venue has one) as a means of escape. If the power goes off you don't want to be stuck in a lift inside a burning building.

To prevent panic amongst the public, announcements are sometimes given in a special code to warn staff and crews of fire and other dangers, such as bombs. Make certain that you know the emergency procedures in place at each given workplace and what your instructions are if such an announcement is made.

When calling the emergency services in the UK, we normally use the 999 system, but this varies in some establishments and in other countries. To bring us in line with the rest of Europe, the emergency telephone number is due to change in the UK in the near future to the standard European fire emergency: 112.

After dialling 999, stay at the phone and ask for the emergency service required (Police, Fire or Ambulance). Nervous tension tends to make our speech speed up, so take a deep breath and speak slowly and clearly giving the precise location of the incident.

Smoke machines and the smoke given off by pyrotechnics may trigger automatic smoke detectors fitted in many venues. Never attempt to cover or block off these alarms so that you can use these effects – manslaughter is a serious charge!

When attempting to escape from a smoke filled room or building, a hands and knees crawling position is best adopted, keeping as low to the ground as possible to avoid the heat.

The mouth and nose should be covered with a cloth such as a scarf or handkerchief to help prevent smoke inhalation.

The palms of the hands should not be placed flat on the ground. Instead, clench the fists and move on your knuckles; if a live electrical cable has become exposed and the flat hand is placed upon it, the shock will cause the hand to close and grip the live cable. However, if the knuckles touch the live cable, the shock will cause the hand to be thrown away from the cable.

Tackling a Fire

Fire prevention is always better than cure, but if there is a fire in your workplace, only tackle this fire if it is safe to do so!

In the workplace you will find a variety of fire-fighting equipment ranging from fire buckets containing water or sand to hose reels to fire blankets, as well as a variety of fire extinguishers.

Types of Fire

Fires are classified an accordance with British Standard 4547 and are defined as follows:

Class A Fires

Fires involving ordinary combustible materials such as wood, cloth and paper

Class B Fires

Fires involving flammable liquids or liquefiable solids (petrol, paraffin, paints), oils greases and fats.

Class C Fires
Fires involving gases.

Class D Fires
Fires involving burning metals.

Class F Fires
Fires involving cooking oils.

Electrical Fires

Types of fire Extinguisher

The main types of fires extinguisher are as follows:

Water (Red)

Suitable for Class A fires – water works by cooling.

Point the jet at the base of the flames and keep it moving across the fire. Ensure the whole fire is out.

Do not use on electrical fires or on burning fat or oil.

Foam (Cream)

Suitable for a limited number of Class B fires – foam mainly works by smothering and cooling. Check the manufacturer's instructions for suitability of use on fires involving liquids.

Point the jet at the base of the flames and keep it moving across the fire. Ensure the whole fire is out. For fires involving liquids, do not aim the jet straight into the liquid. Where the liquid is in a container, point the jet at the inside edge of the container or on a nearby surface above the burning liquid. Allow the foam to build up and flow across the liquid.

AFFF (Aqueous Film-Forming Foam – Cream)

Suitable for Class A and B fires – foam mainly works by smothering and cooling.

Danger – some foam types are not suitable for use on live electrical fires.

Point the jet at the base of the flames and keep it moving across the fire. Ensure the whole fire is out. For fires involving liquids, do not aim the jet straight into the liquid.

Where the liquid is in a container, point the jet at the inside edge of the container or on a nearby surface above the burning liquid. Allow the foam to build up and flow across the liquid.

Standard Dry Powder or Multi-Purpose Dry Powder (Blue)

Suitable for Class B fires.

Safe on live electrical equipment, though does not penetrate spaces inside equipment. A fire may re-ignite as powder does not have a cooling effect and does not penetrate inside spaces. Powder can create a mess on delicate equipment.

Multi-purpose dry powder melts to form a skin that smothers the fire and provides some cooling effect.

Point the jet discharge horn at the base of the flames and, with a rapid sweeping motion, drive the fire towards the far edge until all the flames are out.

If the extinguisher has a shut-off control, wait until the air clears, and if the flames are still visible, attack the fire again.

Carbon Dioxide or CO² (Black)

Suitable for Class B fires.

Safe and clean to use on live electrical equipment. Does not cool the fire very well and you may need to watch that the fire does not restart. Ventilate area after use in a confined space as Carbon Dioxide displaces oxygen in the air to put out the fire and not by cooling.

The discharge horn should be directed at the base of the flames and the jet kept moving across the fire. The discharge horn gets extremely cold while the extinguisher is in use. These extinguishers are also very noisy.

Vaporising Liquid or Halon (Green)

Suitable for Class B fires.

Clean and light. Can also be used on small surface burning Class A fires.

Effective and safe on live electrical equipment.

Due to the ozone-depleting potential of Halon, its future use and availability will be restricted to 'Essential Uses'. Production and consumption of new HALON ceased in 1994. Halon does not cool the fire very well and you may need to watch that the fire does not restart again. Ventilate area after use in a confined space, as fumes from vaporising liquid can be harmful, particularly if used on hot metal.

The vaporising liquid gas is expelled in a jet which should not be aimed into burning liquids as this risks spreading the fire. The discharge nozzle should therefore be aimed at the flames and kept moving across the fire.

Other fire-fighting equipment:

Hose reels (Water)

Suitable for Class A fires.

Danger: Do not use on live electrical equipment.

Point the jet at the base of the flames and keep it moving across the the fire. Ensure all the fire is out.

Fire Blankets

Suitable for Class A and B fires.

Light duty: Suitable for burning clothing and small fires involving burning liquids.

Heavy duty: Suitable for industrial use; it is resistant to penetration by molten metals.

Whatever type or make of fire extinguisher you choose, make sure it conforms to the appropriate British Standards (BS EN3 and BS 7863). Look for the Kite-mark or the special British Approvals for Fire Equipment (BAFE) mark.

- Fire extinguishers should *never* be used on fat pan fires.
- Try to keep your head low, beneath the level of any fumes or vapours.
- Never tackle a fire if it is starting to spread or if the room is filling with smoke. The most common cause of fire deaths is being overcome by smoke or fumes.
- If you cannot put out the fire or if the extinguisher becomes empty, evacuate the area, closing all doors and windows behind you. Do not wait to pick up personal belongings but turn off the electricity or gas supplies if possible – it is safe to do so.
- Do not move an object on fire.
- Always call the fire brigade even if you have managed to put the fire out.
- Do not open a door if it or handle is hot or if smoke is coming out from around the door, unless if it is absolutely essential. If this is the case, stay behind the door so that it acts as a shield should a sheet of flame or a fire ball suddenly shoot out as you open the door.

Basic Fire Safety Training

Basic Fire Safety Training is a legal requirement and must be given to all new staff upon joining a company as part of their induction. The training should include information on:

How Quickly Fire Spreads

Fire Prevention includes electrical safety, chemical safety and the need for good housekeeping by keeping the work place clean and tidy and obeying 'No Smoking' rules. The reasons for not obstructing, locking or propping open fire doors, exits or other means of escape should also be explained.

The reason for not obstructing fire safety information signs, fire alarms, fire-fighting equipment or erecting P.A. stacks or other equipment under the 'Iron Curtain' in a theatre should be explained.

How to raise the alarm: who is responsible? What happens when they are sick or on holiday? Staff should know the particular arrangements in their regular place of work and venues they may visit.

What to do upon hearing the alarm. Switching off equipment and closing windows if time permits; not stopping to collect personal belongings; calmly walking to the nearest exit and assembling at the designated point.

How to tackle a fire in safety: instruction on how to use basic fire-fighting equipment, such as hose reels, fire blankets, water and sand buckets and the different types of fire extinguishers and what type of fire to use them on.

Many Fire Brigades (check with your local Fire Brigade HQ) and commercial operations such as Stagesafe now offer half or one-day fire safety training courses to businesses and other groups. These are well worthwhile and often each participant is given a certificate to prove he/she has received such training. This training includes fire prevention as well as selection and use of basic fire-fighting equipment.

Detailed records must be kept on the fire training given to staff. Records must also be kept of weekly fire-alarm tests, details of fire drills including the time and date the alarm was raised, the time taken to evacuate the premises, names of those who took part in the drill together with any further comments, defects and remedial action. These records should be kept with records of test and inspection for fire safety and fire-fighting equipment such as alarm systems and fire extinguishers.

38 CONTROL OF SUBSTANCES HAZARDOUS TO HEALTH REGULATIONS 2002 (COSHH), CHEMICALS (HAZARD INFORMATION AND PACKAGING) REGULATIONS 2002 (CHIP) AND NOTIFICATION OF NEW SUBSTANCES REGULATIONS 1993 (NONS)

Yes, it's Risk Assessment time again. This time we are looking to see if we have any hazardous substances and chemicals in our workplace and then assessing to see if we are using and storing them in the safest possible way.

The COSHH Regulations have been introduced to protect people who come into contact with hazardous substances as part of their work.

Fortunately, we don't usually use that many very hazardous substances in our industry, but there are quite a few with a lower hazard rating. The legal requirement is for the assessment to be "suitable and sufficient" (as it is with all assessments). More serious and complex risks require greater consideration to meet this requirement; simpler and lower risks will require less. COSHH defines substances as any natural or artificial substances, whether in solid or liquid form, or in the form of a gas or vapour.

'Hazardous substances' include:

- Chemicals classified as 'very toxic', 'toxic', 'harmful', 'irritant' or 'corrosive'.
- Any substance which has been assigned a maximum exposure limit or occupational exposure standard.
- Substantial concentrations of airborne dust.
- Harmful micro-organisms.
- Any other substance which could harm people's health.
- The main duty that COSHH places on employers is to conduct a

risk assessment before any work involving hazardous substances is undertaken.

Chemicals that are classed as explosive, oxidising, extremely flammable or highly flammable are classed as a physicol chemical hazard and are not covered by COSHH and radioactive substances emitting ionising radiation are covered by their own regulations and again are not covered by COSHH.

Caterers will probably find they have more to do in terms of assessments as they use a large number of cleaning substances. However, nickel is classed as a hazardous substance and, as any sound engineer will tell you, microphones contain nickel. The first step is to decide if there are any hazardous substances in the workplace. If the answer is 'No', then no further action is required. If it is 'Yes', then an assessment is required. It may be possible to replace a high-risk substance with a lower risk one.

To aid this process, product manufacturers have a duty under the Chemicals (Hazard Information and Packaging) Regulations 2002 (CHIP) to supply Safety Data Sheets with their products.

CHIP requires suppliers to identify the hazards (or dangers) of the chemicals they supply. This is called classification. If a chemical is dangerous, your supplier (who may be the manufacturer, importer or distributor) must provide you with the information about the hazards the chemical presents; you should not receive the safety information (Safety Data Sheet) later than the product itself. However, it may be sent separately, by fax or email, for example.

Safety Data Sheets give details of the chemical composition of the product, correct storage conditions, the properties of the product, i.e. toxic if swallowed, flammable, etc., correct use and what to do in an emergency.

CHIP further requires suppliers to ensure all chemicals they supply are properly packaged and correctly labelled.

The HSE also produces an Approved Supply List that provides information approved for the classification and labelling of substances and preparations dangerous for supply. This approved supply list is extremely useful for anyone involved is COSHH assessments.

Read the Label

Most hazardous products have a warning and hazard marking on the container, but an important new requirement of CHIP is that your supplier must provide you with more detailed hazard information on a *Safety Data Sheet.*

Safety data sheets are important in helping you, or anyone you supply, to make the workplace safe and protect the environment.

More specifically, a Safety Data Sheet contains information to help you make a risk assessment as required by COSHH. You will almost certainly need a Safety Data Sheet and you will need to read the label on containers of substances as well as consult the HSE Approved Supply List before you can carry out your assessment.

The Safety Data Sheet and the Approved Supply List are not assessments. However, they will describe the hazards of a chemical, helping you to assess the probability of those hazards (i.e. the risk) arising in the workplace.

CHIP does not cover all hazardous chemicals; some groups such as pesticides, cosmetics and medicines are covered by other legislation and rules for packaging and labelling.

Retailers do not have to supply Safety Data Sheets to the general public, but if you buy a dangerous chemical from a retailer for use at work, the retailer must provide a Safety Data Sheet if you ask for one. They do not have to give you the data sheet with the product provided they make arrangements to forward it promptly.

Remember, Safety Data Sheets only have to be provided with dangerous chamicals. If a chemical is not classified as dangerous under CHIP, it does not need a Safety Data Sheet.

Safety Data Sheets should be kept with the substances they cover. A good idea is to laminate them and pin them to the door of the room in which the substances are stored.

Substances may be classed as toxic, flammable, corrosive, explosive, irritants or harmful. Employers must instruct staff on the safe use of hazardous substances and warn them of the dangers associated with their use, such as fire (smoking) inhalation of fumes and gases, contact with the eyes or skin, etc. You may find a substance marked with a Maximum Exposure Limit (MEL) or a MEL on a safety data sheet; this is the maximum dose allowed in a 15-minute period. A full assessment is required for such a substance and a copy of the assessment must be given to each operative who works with such substances, together with the required training and Personal Protective Equipment/Clothing.

An Occupational Exposure Standard (OES) is for a standard eight-hour working day. The HSE publishes lists of Occupational Exposure Standards and Maximum Exposure Limits for a whole range of substances, which may be required to complete risk assessments.

I have obtained Safety Data Sheets for several different (but not all) types of Smoke Fluid, Cracked Oil and Flambar. The Safety Data Sheets state that these substances do not carry a Hazard Rating and that in normal use it will be almost impossible to get anywhere near a MEL or OES with these products. However, you are advised to wear a respiratory mask when using large quantities of Flambar (See Personal Protective Equipment Regulations).

A respiratory mask will also protect against the dust (another form of hazardous substance) that comes off drapes when they are unfolded. It must be noted that not all of these substances are free of Hazard Ratings, so always check the products you are using and never assume.

A lot of controversy has surrounded the use of MDF (Medium Density Fibreboard) that is made from recycled residues and by-products from the forestry and saw milling industry. Rumours have claimed that MDF is highly toxic and is banned in many European countries; this is not true. MDF is not banned in any European countries and the HSE states that there is no evidence to suggest that MDF is any more toxic than any other type of wood.

Assessments are required when cutting, drilling or machining any kind of wood including MDF, since the dust can be a problem. Dust extraction systems should be employed and dust masks worn by operators. Dust suspended in the atmosphere can be highly explosive and a fire hazard. The HSE has published guidance sheets on the hazards associated with wood and wood dust, which should be referred to for further information.

Employers must warn employees never to mix substances and to always store the substances in correctly labelled containers suitable for the substance, which is normally the container the substance was originally supplied in. Substances must be stored in a secure place in accordance with the manufacturer's instructions, safety information on the label, or Safety Data Sheets.

Further instruction must be given to employees not to dispose of substances into sinks, drains or watercourses, as this may result in pollution. They should also be instructed not to dispose of substances or containers on fires as this may also result in pollution and/or the risk of explosion.

The substances listed below can be found at most of the workplaces we are involved with, i.e. venues, concert halls, clubs, arenas, stadiums and festival sites. I have not included warehouses or manufacturing units. In most cases they are the type of substance that may be found in most offices and domestic situations.

Dry ice, gas and diesel need special consideration. LPG is covered by the

Highly Flammable Liquids and Liquefied Petroleum Gases Regulations 1972 and the Gas Safety Installation and Use Regulations 1998. Provided these substances are used in well-ventilated environments, the precautions mentioned, along with the use of relevant PPE (aprons, gloves, face-masks), should prove sufficient requirements for use of dry ice, gas and diesel.

Special attention is needed in the preparation of gobos as strong chemicals and acids are often used and these substances will need detailed assessments.

Employees should be instructed to:

- Use the required equipment (including PPE).
- Clean up spillages and leakages promptly and dispose of any excess caused by the spillage or leakage in the approved way (according to the Safety Data Sheet).
- Never dispose of chemicals or waste in streams or water courses.
- Never pour chemicals down the sink or dispose of them on a fire, especially aerosol containers.
- Wash their hands and other exposed parts of the body after using any substances.

In Catering, Office and Toilet Accommodation:

Photocopier/fax toner, typist correction-fluids, inks, glue, ozone generated from photocopiers, bleach, urinal blocks, toilet cleaner, organisms, liquid and solid waste, window cleaner, furniture polish, floor cleaner, disinfectant, sink cleaner, washing-up liquid, washing powders, oven cleaner, fabric softener/conditioner/freshener, gas, fly-sprays and other pest control substances, air freshener, cigarette smoke, ash and dust.

The suppliers of mobile and portable toilet units will have a few assessments to carry out, and I don't just mean 'going through the motions'!

On and Around Stage Areas:

Paints, adhesives, solvents, varnish, water-proofing substances, electrical spray, fireproofing substances, exhaust fumes, dust, fluids for smoke and vapour effects units, dry ice, oil, grease and diesel. (Petrol should have no place in our work apart from in our cars.)

Having decided that hazardous substances do exist in our workplace, an employer must carry out an assessment. This needs to be done by a 'competent' (trained) person. At the end of each stage of the assessment, ask yourself if you are genuinely confident about the assessment so far. If not, go back over what you have done. The following steps are involved in the assessment:

1. Gathering Information about the Substances, the Work and Working Practices

- This involves finding out what substances are present or likely to be present
- Identifying the hazards they have
- Finding out who could be exposed and how, *either* by seeing which substances occur in particular activities *or* by seeing which activities involve exposure to particular substances

2. Evaluate the Risks to Health

Either on an individual employee basis *or* on a group basis.

Find out:

- The chance of exposure occurring.
- What level of exposure could happen
- How long the exposure lasts.
- How often exposure is likely to occur

Conclude:

- Either existing and potential exposure poses no significant risk *or* existing and/or potential exposure poses significant risk

3. Decide what needs to be done in terms of

- Controlling or preventing exposure
- Maintaining controls
- Using controls
- Monitoring exposure
- Health surveillance
- Information, instruction and training

4. Record the Assessment

- Decide if it is necessary to record the assessment
- If YES, decide what and how much to record
- Decide presentation and format

5. Review the Assessment

- Decide when review is needed

- Decide what needs to be reviewed

The hierarchy of controls to reduce exposure are:

- Using appropriate work processes, systems and engineering controls, and providing suitable work equipment and materials.
- Controlling exposure to hazardous substances at source. For example, by introducing adequate ventilation systems and appropriate organisational measures in the workplace.
- Providing suitable personal protective equipment where adequate control cannot be achieved by other means.

The regulations also set out additional duties for employers. For example, a requirement to draw up detailed procedures for dealing with accidents, incidents and emergencies involving hazardous substances in the workplace, including carrying out regular safety drills and ensuring that only authorised personnel are allowed to enter danger areas in the event of an emergency.

In addition, the regulations require employers to keep up-to-date records of individual workers who undergo health surveillance under the regulations.

The ACoP includes an appendix to help employers protect their employees from exposure to asthma-causing substances at work, including factors that must be taken into account when carrying out risk assessments under COSHH.

It cannot be over-emphasised that the depth of assessment required depends on the complexity and degree of risk. Simple, low-risk situations will require little, but high-risk complex situations need much more attention. It might take one person two minutes to assess the risks from using correction fluid in the office, yet it could take multi-disciplinary team weeks to assess the risk in the factory where it was made.

As with all assessments, the regulations require assessments to be reviewed. This means re-examining earlier conclusions, but if those earlier conclusions are still valid it is not necessary to repeat the whole process. The primary purpose of review is to check and, where necessary, amend assessments, not repeat them.

I said earlier that I was not going to mention warehouses or manufacturing units, but I should just mention that during my travels I have found various chemicals in PA and lighting company warehouses that are used for etching printed circuit boards and gobos. Most of these chemicals are corrosive and include sodium persulphate, metasilicate and ferric chloride hexahydrate. I don't know much about these substances but I do know that Safety Data Sheets and Risk Assessments are required!

39 DANGEROUS SUBSTANCES AND EXPLOSIVE ATMOSPHERES REGULATIONS 2002 (DSEAR)

The Dangerous Substances and Explosive Atmospheres Regulations 2002 (DSEAR) came into force on 9th December 2002. The regulations revoke the Highly Flammable Liquids and Liquid Petroleum Gas Regulations 1972, amongst others. DSEAR applies to all dangerous substances at nearly every business in the UK. It sets minimum requirements for the protection of workers from fire and explosion risks arising from dangerous substances and potentially explosive atmospheres. DSEAR complements the requirement to manage risks under the Management of Health and Safety at Work Regulations 1999. A brief summary of the main requirements of DSEAR is given below.

Employers and the self-employed must:

- Carry out a risk assessment of any work activities involving dangerous substances.
- Provide technical and organisational measures to eliminate or reduce as far as is reasonably practicable the identified risks.
- Provide equipment and procedures to deal with accidents and emergencies.
- Provide information and training to employees.
- Classify places where explosive atmospheres may occur into zones, and mark those zones where necessary. (This duty is being phased in – see table on page 332 for dates).

Overall, DSEAR can be seen as complementary to the general duty to manage risks under the Management of Health and Safety at Work Regulations 1999, making explicit good practices for reducing the risk to persons from fires, explosions and similar energetic (energy-releasing) events which are, in turn, caused by dangerous substances such as flammable solvents. The impact of DSEAR on the diligent employer should therefore be small. Other than for certain maritime activities, DSEAR applies whenever the following conditions have been satisfied:

1. Work is being carried out by an employer or self-employed person.

2. A dangerous substance is present or is likely to be present at the workplace.

3. The dangerous substance presents a risk to the safety of persons (as opposed to a risk to health).

The definition of 'workplace' is a general term for any premises or part of premises used for work. Premises include all industrial and commercial premises. It also includes land-based and offshore installations as well as vehicles and vessels. Common parts of shared buildings, private roads and paths on industrial estates, and business parks are also 'premises' – as are houses and other domestic dwellings.

If there is a work activity in 'premises' as defined above, then it is a workplace for DSEAR purposes.

However, the requirements in DSEAR concerning zoning and the coordination of safety measures in shared workplaces do not apply to all workplaces. Certain sectors and work activities are exempt because there is other legislation fulfilling these requirements, for example, in the offshore sector.

DSEAR applies to any substance or preparation (mixture of substances) with the potential to create a risk to persons from energetic events such as fires, explosions, thermal runaway from exothermic reactions etc.

Such substances, known in DSEAR as dangerous substances, include: petrol, liquefied petroleum gas (LPG), paints, varnishes, and certain types of combustible and explosive dusts produced in, for example, machining and sanding operations.

It should be noted that many of these substances can also create a health risk. For example, many solvents are toxic as well as flammable. DSEAR does not address these health risks. They are dealt with by the Control of Substances Hazardous to Health Regulations (COSHH), which have been amended.

DSEAR is concerned with harmful physical effects from thermal radiation (burns), over-pressure effects (blast injuries) and oxygen depletion effects (asphyxiation) arising from fires and explosions.

The steps necessary to determine whether you have a dangerous substance present in your workplace are set out below.

The following examples illustrate the type of activities and substances commonly found at work that are likely to be covered by DSEAR:

- Storage of petrol as a fuel for cars, motorboats, horticultural machinery, etc.

- Use of flammable gases, such as acetylene, for welding.
- Handling and storage of waste dusts in a range of manufacturing industries.
- Handling and storage of flammable wastes, including fuel oils.
- Hot work on tanks or drums that have contained flammable material.
- Work activities that could release naturally occurring methane.
- Dusts produced in the mining of coal.
- Use of flammable solvents in pathology and school laboratories.
- Storage/display of flammable goods, such as paints, in the retail sector. Filling, storage and handling of aerosols with flammable propellants, such as LPG.
- Transport of flammable liquids in containers around thc workplace.
- Deliveries from road tankers, such as petrol or bulk powders.
- Chemical manufacture, processing and warehousing.
- Petrochemical industry - onshore and offshore.

DSEAR is intended to protect not only employees at the workplace, but also any other person, at work or not, who may be put at risk by dangerous substances. This includes employees working for other employers, visitors to the site, members of the public, etc. However, when considering arrangements to deal with accidents, incidents and emergencies and the provision of information, instruction and training, employers only have duties to persons who are at their workplace. DSEAR requires employers (or self-employed persons) to:

1. Carry out a risk assessment before commencing any new work activity involving dangerous substances.
2. In the case of an employer with 5 or more employees, to record the significant findings of the assessment as soon as is practicable after the assessment is made, including:
   - The measures (technical and organisational) taken to eliminate and/ or reduce risk.
   - Sufficient information to show that the workplace and work equipment will be safe during operation and maintenance including:
   - Details of any hazardous zones.
   - Any special measures to ensure coordination of safety measures and procedures, when employers share a workplace (a. and b. apply from 1 July 2003).

- Arrangements to deal with accidents, incidents and emergencies.
- Measures taken to inform, instruct and train employees.

The risk assessment required by DSEAR is an identification and careful examination of the dangerous substances present in the workplace, the work activities involving those substances and how they might fail dangerously, giving rise to fire, explosion and similar events with the potential to harm employees and the public. Its purpose is to enable employers to decide what they need to do to eliminate or reduce, as far as is reasonably practicable, the safety risks from dangerous substances. The risk assessment is required to be carried out before commencing any new work activity and DSEAR also requires that the measures identified as necessary by the risk assessment are implemented before the work commences. Employers are required to ensure that the safety risks from dangerous substances are either eliminated or reduced as far as is reasonably practicable. Where it is not reasonably practicable to eliminate risks, employers are required to take, so far as is reasonably practicable, measures to control risks *and* measures to mitigate the detrimental effects of a fire, an explosion or a similar event. Hence DSEAR reflects the well-understood safety hierarchy of:

- Elimination or
- Control and Mitigation.

Elimination is the best solution and involves replacing a dangerous substance with a substance or process that fully eliminates the risk. In practice this is difficult to achieve and it is more likely that it will be possible to replace the dangerous substance with one that is less hazardous (e.g. by replacing a low flashpoint solvent with a high flashpoint one). Another option would be to design the process so that it is less dangerous, for instance by reducing quantities of substances in the process, known as process intensification. However, care must be taken whilst carrying out these steps to ensure that no other new safety or health risks are created or increased.

DSEAR requires that control measures are applied in the following priority order consistent with the risk assessment and appropriate to the nature of the activity or operation. DSEAR requires that mitigation measures consistent with the risk assessment and appropriate to the nature of the activity or operation are applied including:

- Reducing the number of employees exposed.
- Providing plant which is explosion-resistant.

- Providing explosion suppression or explosion relief equipment.
- Taking measures to control or minimise the spread of fires or explosions.
- Providing suitable Personal Protective Equipment (PPE). DSEAR also specifies that the measures taken to achieve the elimination or the reduction of risk should include:
- Design, construction and maintenance of the workplace (e.g. fire-resistance, explosion relief).
- Design, assembly, construction, installation, provision, use and maintenance of suitable work processes, including all relevant plant, equipment, control and protection systems.
- The application of appropriate systems of work, including written instructions, permits to work and other procedural systems of organising work.

DSEAR also requires the identification of hazardous contents of containers and pipes. Many will already be marked or labelled under existing EC legislation. For those which are not, 'identification' could include training, information or verbal instruction, but some may require labelling, marking or warning signs. In workplaces where explosive atmospheres may occur, you should ensure that:

- Areas where hazardous explosive atmospheres may occur are classified into zones based on their likelihood and persistence.
- Areas classified into zones are protected from sources of ignition by selecting equipment and protective systems meeting the requirements of the Equipment and Protective Systems Intended for Use in Potentially Explosive Atmospheres Regulations 1996, although equipment in use before 1st July 2003 can continue to be used indefinitely, provided the risk assessment shows it is safe to do so.
- Where necessary, areas classified into zones are marked with a specified 'EX' sign at their points of entry.
- Where employees work in zones areas they are provided with appropriate clothing that does not create a risk of an electrostatic discharge igniting the explosive atmosphere.
- Before coming into operation for the first time, areas where explosive atmospheres may be present are confirmed as being safe (verified) by a person (or organisation) competent in the field of explosion protection. The person carrying out the verification must be

competent to consider the particular risks at the workplace and the adequacy of control and other measures put in place. These additional requirements come into effect at different times, depending on when the workplace is first used.

Workplace	When requirements must be met
Workplace in use before July 2003	Workplace must meet requirements by July 2006
Workplace in use before July 2003 but modified before July 2006	Workplace must meet requirements from the time the modification takes place
Workplace coming into use for the first time after June 2003	Workplace must meet requirements from the time it comes into use

This part of DSEAR complements the Equipment and Protective Systems Intended for Use in Potentially Explosive Atmospheres Regulations 1996. DSEAR requires that employers make arrangements to protect employees (and others at the workplace) in the event of accidents, etc. The provisions build on existing requirements in Regulation 8 of the Management Regulations, and require employers to make arrangements including:

- Suitable warning (including visual and audible alarms) and communication systems.
- Escape facilities – if required by the risk assessment.
- Emergency procedures to be followed in the event of an emergency.
- Equipment and clothing for essential personnel dealing with the incident.
- Practice drills.
- Making information on the emergency procedures available to employees.
- Contacting the emergency services to advise them that information on emergency procedures is available (and providing them with any information they consider necessary).

The scale and nature of the emergency arrangements should be proportionate to the risks. These requirements clarify what already needs to be done in relation to the safety management of dangerous substances, and will not require any duties in addition to those already present in existing legislation. Employers are required to provide employees and others at the workplace who might be at risk, with suitable information, instruction and training on precautions and actions they need to take to safeguard themselves and others, including:

- Names of the substances in use and risks they present.
- Access to any relevant safety data sheet.
- Details of legislation that applies to the hazardous properties of those substances.
- The significant findings of the risk assessment.
- The significant findings of the risk assessment.

Employers should also make information available to employee representatives. Information, instruction and training need only be provided to non-employees where it is required to ensure their safety. Where it is provided, it should be in proportion to the level and type of risk. Much of this is already required by existing health and safety legislation. Petroleum legislation is being modernised as part of the DSEAR package.

Previously, the keeping of petrol was controlled by licenses issued under the Petroleum (Consolidation) Act 1928. However, the new DSEAR Regulations apply to petrol and therefore duplicate these controls. Accordingly, DSEAR remove licensing requirements for petrol, except for petrol kept for dispensing into vehicles (retail and non-retail). Work is continuing to further modernise the petrol regime. Determination of the presence of dangerous substances You will need to carry out the following two steps:

1. Check whether the substances have been classified under the Chemicals (Hazard Information and Packaging for Supply) Regulations 2002 (CHIP 3) as explosive, oxidising, extremely flammable, highly flammable or flammable.
2. Assess the physical and chemical properties of the substance or preparation and the circumstances of the work involving those substances to see if a safety risk to persons could be created from an energetic event.

Step 1: When dangerous substances are used at work, suppliers must provide you with safety data sheets and the safety data sheet should tell you whether the chemical is classified under the CHIP Regulations as flammable, oxidising, etc. Another source of information is the HSE's Approved Supply List. This is a list prepared by the HSE, listing many commonly used substances and their classification. If a substance or preparation is classified as explosive, oxidising, extremely flammable, highly flammable or flammable, then it is a dangerous substance.

Step 2: You will need to carry out a risk analysis using information about the chemical and physical properties of the substance and the circumstances of the work to determine whether a dangerous substance is present. The key point here is that it is the combination of the properties of the substance and the circumstances of the work process that need to be assessed. For example, diesel (or other high flash point) oils are not classified as 'flammable' under CHIP, yet if they are heated to a sufficiently high temperature in a process, they can create a fire risk. In these circumstances the diesel oil becomes a dangerous substance for the purposes of DSEAR. On the other hand, if diesel oil is only present in storage at ambient temperatures, it is not a dangerous substance for DSEAR purposes. Other examples include substances which decompose or react exothermically when mixed with certain other substances, e.g. peroxides. Wood, flour and many other dusts are, depending on the circumstances of the work, also dangerous substances for DSEAR purposes. This is because when the dust is mixed in a cloud with air it can, in certain circumstances, be ignited and explode. Work activities involving grinding or machining are particularly prone to this risk. If the assessment shows that there is a safety risk to persons arising from a fire, explosion or other energy-releasing events, then the substance is a dangerous substance for DSEAR purposes.

As indicated, DSEAR has now revoked the Highly Flammable Liquids and Liquid Petroleum Gas Regulations 1972 Caterers and those who use forklift trucks powered by LPG are probably the only two groups of people working in our industry who are going to be affected by this and the Gas Safety (Installation and Use) Regulations 1998.

LPG is commonly referred to as propane (cylinders are conventionally coloured red) or butane (conventionally blue). It is pressurised into liquid form for storage in cylinders.

The gas forms flammable mixtures with air in concentrations of between 2% and 10% by volume. It is naturally colourless and odourless, although it is usually odorised to enable easy detection at low concentrations. It is heavier than air, and any gas released will tend to accumulate in cellars, pits or drains. It is non-toxic, but will displace oxygen and could therefore cause asphyxia in substantial concentrations.

The physical hazards arising from LPG are fire or explosion resulting from escape and subsequent ignition of gas or the cylinder becoming involved in a fire. Contact with LPG liquid can cause severe cold burns.

In practice, cylinders should be kept outside, away from entrances, exits and circulation areas, and secured in an upright position. They must also be equipped with pressure relief valves.

Combustible materials must be at least 1.5 metres away from cylinders. Think of the implications of this on a festival campsite where there may be hundreds of campers all using small LPG camping stoves in small nylon (highly inflammable) tents and/or close to vehicles (remember petrol, oil and nylon make Napalm), and now do a risk assessment that is suitable and sufficient!

Gas equipment should be checked by a qualified (CORGI) gas fitter who is authorised to service LPG equipment. This should be done at least once a year.

Fixed piping should be used to connect cylinders to gas appliances if possible. If this is not possible, flexible tubing should be to the appropriate B.S. Standard and provided with mechanical protection to minimise damage. Tubing must be secured by crimping, jubilee clips or similar, to make them gas-tight. Gas supplies must be isolated at the cylinder as well as the appliance when not in use.

When replacing gas cartridges they should be fitted in the open air away from sources of ignition.

Catering areas (particularly those set up in tents or marquees) should be specifically laid out for the instillation of cooking equipment. All appliances must be fixed on a firm non-combustible heat-insulating base and surrounded by shields of similar material on three sides. The shields must be at least 600mm (2ft) away from combustible material and care should be taken to prevent combustible material blowing against the cooking equipment.

Further information on the safe use and storage of LPG, including various codes of practice, can be obtained from the Liquid Petroleum Gas Association.

Trucking companies should note that if they transport cylinders of LPG for caterers, then a warning sign must be in place on the truck. However, the regulations regarding the Carriage of Dangerous Goods is scheduled to change and a licence may soon be required by anyone who intends to carry even a single small camping gas cylinder. This will apply to even transporting a cylinder in a private car, and trucks will have to be specially equipped to carry catering gas cylinders.

Basic LPG Safety

- Light the match or igniter before turning on the gas.
- Check to see if the pilot light or main burner has ignited – especially inside ovens where it is not always easy to see the flame.

- If the gas will not ignite easily turn off the gas and check that there is LPG in the supply cylinders. If in doubt, call a gas dealer.
- Remember that LPG vapour is heavier than air and even a small leak will result in gas accumulating on the floor and forming a flammable mixture with the surrounding air.
- Remember, too, that LPG vapour is invisible, but you can quickly detect its presence by its strong smell. Extinguish all flames and do not smoke. Ventilate the area by opening doors and windows until the smell has gone.
- Gas leaks are caused by accidentally leaving open a gas valve or by a faulty connection to a pipe or valve. To find the leak, splash the suspect part of pipe or valve with soapy water. The leak will cause bubbles.
- Cylinders must always stand upright.
- Do not use a cylinder which is damaged, e.g. badly rusted or dented, cut, bulging, etc. – have it checked by your LPG Dealer.
- Only use a proper LPG hose to connect your stove or other LPG appliances. Ordinary rubber or garden hose must never be used, as these are not designed for LPG and will soon deteriorate and leak.
- Keep flexible piping away from heat.
- Flexible pipe should be inspected annually for leaks.
- Flexible piping should be clamped to correctly designed metal connectors.
- A good supply of oxygen (air) is essential for efficient combustion.

Therefore, a room in which LPG is burned must have adequate ventilation openings – at low level to let fresh air in and at high level for products of combustion to exit. This is particularly necessary for small rooms.

- It is important to note that inadequate ventilation may result in the formation of poisonous carbon monoxide.
- Most gas connections utilise synthetic rubber joints or O-rings. These should be inspected (whenever the joint is made or broken) for cracks, perishing or other damage (e.g. brittleness due to ageing), and replaced if necessary.
- If there is a gas leak (LPG or mains gas) that can't be controlled by turning off the supply, then it will be necessary to evacuate the premises using the fire evacuation procedures. I should not need to mention the following requirements: No smoking, no naked lights,

and no operating of electrical switches in this situation.

If the leak is inside a premises then the windows and doors should be opened if possible. If a gas explosion has taken place, any gas flames should be left to burn as this will prevent a further build up of gas.

- Emergency drill for fire at cylinder/s.
- Don't panic – Flames from joints near cylinder are not dangerous in themselves.
- If possible, close the cylinder valve using a wet cloth to protect your hands.
- Spray cold water onto cylinders exposed to the fire in order to keep them cool. Use a hosepipe. Keep people away from the area.

Note: If in doubt about any of the above, contact your LPG dealer or appliance supplier who should be registered with CORGI.

Question:

What is the main cause of injury to electricians at work? The true answer is falling off ladders!

40 CROWD MANAGEMENT AND SAFETY STEWARDS

Stewards have a major and visible role in maintaining and managing safety arrangements at events. An invaluable aspect of this role is the ability and opportunity to act as the eyes and ears of safety officers and advisers as well as the crowd manager. The author prefers to use the term Safety Steward as this term is less confrontational than the word 'security' and in most situations is more appropriate and descriptive of the role undertaken. This chapter only touches on the subject of crowd management; to do it justice would require another book, but it is a very important subject that forms a major part of event safety.

The most readily identifiable problem with the word 'security' is that it often conjures up an image either of 'Rent A Thug' or a bouncer more intent on starting a 'punch up', involved with 'public order' offences, rather than providing a trained professional safety presence with a commitment to 'public safety'; the latter being the role of the safety steward in today's society.

We must first understand that there is a huge difference between crowd management and crowd control. The police understand crowd control but do not seem to understand what crowd management is about. The difference is that crowd management is proactive and involves an understanding of the way crowds form, as well as crowd dynamics. Crowd control is simply reactive, usually when a situation is out of control. This is rather like shutting the stable door after the horse has bolted or trying to cure rather than prevent.

Historically, production crews have displayed a tendency towards showing little or no respect to safety stewards. This attitude is outdated and must change if we are to move forward with a new and modern agenda. We must show stewards more respect, and we can start this process by not jumping over the pit barrier as a shortcut to and from the front-of-house position. When punters see us jumping the barrier, they assume they can do the same and this makes the steward's job tougher. Production staff, artists and guests can also help the process by wearing passes and wristbands. Wristbands must be worn on the wrist not on a lanyard around your neck, so that they are distinct and easily identifiable. Another outdated practice that must be discouraged is the practice of smuggling people backstage and 'blagging' extra passes.

The obvious way in which stewards fulfil their role is by monitoring crowds, checking passes, only allowing access to safety critical areas to authorised persons and vehicles, by preventing banned or illegal articles being brought onto the site or into the venue and by directing traffic, controlling speed limits and restricting traffic movement on outdoor sites. Stewards can check on the state of the toilets and the build-up of litter, waste and combustibles, and generally act as walking 'information points' and as 'banks men' for plant operators. This does not mean that they have to deal with and solve all these problems themselves, rather they should report such incidents and problems so they can be promptly dealt with by the appropriate person or persons. Stewards also have a major role to play in the event of a fire or some other kind of serious incident such as a bomb scare that will require their skills in an evacuation of the venue or event site.

In the foreword to my first book Steve Anderson, Environmental Health Manager for Mendip District Council with responsibility for licensing Glastonbury Festival, made the following comments: *"Health and Safety is a serious business and warrants serious consideration, not just in the more traditional aspects but also in relation to crowds, crowd management and event stewarding – an area that warrants more attention at the moment than most others."* It's taken a long time but our industry is now well on target for setting the standards for event stewards, thanks to the work of the UK Crowd Management Association.

The UK CMA is a professional forum working on setting mutually agreed national standards, including the introduction of formally recognised qualifications by an awarding body called NCFE and another similar qualification by SITO (Security Industry Training Organisation). These two qualifications underpin the new British Standard Code of Practice (BS8406: 2003) for Event Stewarding and Crowd Safety Services.

It must be noted that the new Code of Practice is for Crowd Stewarding and does not include door supervisors or security work such as guarding. In 2002 a new government (cabinet office) watchdog organisation was established (under the Private Security Industry Act 2001) in the form of The Security Industry Authority (SIA). The SIA has already introduced a new standard and qualification for Door Supervisors that became a compulsory national standard in 2004. The new qualification is operated by NCFE, City and Guilds and the British Institute of Innkeeping.

After undergoing training and gaining the required qualification the SIA grants Licences to approved individuals. The SIA also operate an approved contractor scheme that (hopefully) will ensure training, recruitment, supervision and quality systems conform to SIA guidelines and criteria. Checks are carried out by the SIA with the Criminal Records Bureau on criminality criteria for all individuals applying for a licence, a positive move to make the security industry more professional and get rid of the 'cowboys'.

The Private Security Industry Act also makes it a requirement for event stewards working at premises licensed under the 1964 Licensing Act who carry out the following duties to be trained and licensed by the SIA:

a) Guarding premises against unauthorised access, disorder or damage
b) Guarding property against theft, appropriation or damage
c) Guarding people against assault or injuries from unlawful conduct

This requirement is to be extended to premises with other types of licence including places licensed by the Private Places of Entertainment (Licensing) Act 1967, the Local Government (Miscellaneous Provisions) Act 1982, the London Government Act 1963 and Licensing Act 2003.

This potentially means that an event steward may require three different licences (for Event Stewarding, Security (or Static) Guarding and Door Supervision) before he can carry out his duties. The UK CMA are currently engaged in dialogue with the SIA in an attempt to solve these and other potential problems but so far the dialogue seems to be one-sided; the SIA listen but the legislation seems to make it impossible for them to act in a manner in which positive changes can be made.

Crowd marshals and box office staff who only carry out basic tasks such at selling or checking tickets, invitations and passes, together with customer care duties, will not require a licence. The requirement for a licence will be dictated by what duties are in fact carried out and not by what job title is given.

The cost of training and a licence valid for three years is in the region of £450 and the licence is given to the individual, not the stewarding company. Stewarding companies are going to face a tough time trying to find the huge costs involved in finding the finance to train and license staff and then keeping those staff – as once licensed they can then find more lucrative employment. Because of the nature of the work, the Inland Revenue insist that event stewards are part time PAYE employees of stewarding companies (usually working close to the minimum wage), not only does the stewarding company have to pay for the cost of training and licenses they will also have to pay the wages for a steward during the four day training period – that is if the candidate obtain the time off from their regular employment.

Like the 2003 Licensing Act, the introduction of the Private Security Industry Act has created new problems. It remains to be seen but it appears that it is ill-conceived piece of legislation with little or no consideration to the requirements of event stewarding industry. The industry has made it known that it welcomes licensing and will cooperate and respond with a positive commitment and input to develop standards and legislation, but proper consultation has not been carried out to prevent inappropriate licensing or conditions being introduced.

Progress is being made following the introduction of the new qualifications and BS Code of Practice, however much remains uncharted and there are still many outstanding areas needing further attention. The new qualifications have all been specifically introduced for stewards. No national qualifications or training courses exist as yet for steward supervisors, crowd safety managers or event control room staff and as we have seen, it is vital to understand that health and safety arrangements need to be fully embedded at leadership level. This means that the safety message must be communicated from the top down. This situation is also likely to change very soon as key industry figures from the UK CMA and event industry are working with a University College to develop and operate a new foundation degree qualification scheme for crowd managers.

As a by-product of the foundation degree course, training and qualifications for supervisors and pit teams will be produced and will come on stream as 'stand-alone' qualifications. In fact, training course material has already been written.

A recognised problem area that needs specific attention and prioritisation for crowd safety managers and staff is the distinct cultural behaviour that has developed within certain types of audience profile. I am of course referring to the growing propensity and popularity of moshing, stage diving and crowd surfing. Of the three, crowd surfing needs to be prioritised for immediate pre-emptive action. Stage diving can be minimised by a combination of architectural, physical and managerial controls, and this is now beginning to be addressed.

Some work is already being undertaken on crowd surfing and moshing, with the introduction of methods of providing education and information to fans. This is currently being accomplished by the distribution of leaflets and the erection of signs at events spelling out the dangers of crowd surfing and moshing by highlighting the possible consequences. As an additional measure that backs up this policy, any fans who persist with crowd surfing or moshing, despite warnings, may be ejected from an event – three strikes and you're out!

If a stewarding company is required to provide security services (this usually involves asset protection) then this must be agreed and included within the statement of intent.

Those involved in security duties will require additional training and licensing by the SIA. Stewarding companies will require additional types of insurance to cover security work and the client will need to provide an inventory of equipment/assets to be protected. Many claims for lost items of equipment have been made from stewarding companies who were usually unaware if the 'missing' items even existed in the first place. All security work should be carried out in accordance with BS 7499.

These changes are long overdue but do beg the question: "How many promoters will be prepared to pay for fully trained stewards and companies run to British Standards?"

It was thought that unless Entertainment Licence conditions required stewarding companies to operate according to BS8406:2003 for stewards to be and trained to NCFE or SITO standards then the danger remains that promoters will tend to go for a 'cowboy' stewarding company as a cheaper option. However, the Private Security Industry Act has circumnavigated that problem but we will have to wait to see if we have been dealt the best hand.

We will also have to wait to see what is in store with the Security Industry Authority as it will almost certainly be they who call the shots in future. But, we must remember that they are only an enforcement and watchdog agency, they have no powers to change the law (that exists in the form of the Private Security Industry Act). Stewarding companies who wish to be audited and assessed for compliance with British Standards 8406: 2003 and/or quality management standards such as ISO 9001: 2000, should contact the National Security Inspectorate who carry out these duties and issue the quality awards.

When selecting a stewarding company, like any other contractor they should have:

- Public Liability, Employers' Liability and any other insurances as required.
- A Health and Safety Policy.
- Risk Assessments.
- A Code of Conduct.
- A Policy for dealing with children and young persons.
- A Dress Code that includes a means of individual identification for

each steward, such as a number that must be displayed at all times.

Full details of staff training must be available that include:

- Customer care and social skills.
- Duties of an event steward.
- First Aid training and drug awareness (Stewards should be considered and consider themselves as 'outreach workers' for medical and welfare organisations).
- Radio and Communication training (including Social and Communication skills).
- A basic knowledge of Civil and Criminal Law.
- Entertainment Licensing legislation.
- Health and Safety legislation.
- Emergency Planning and Evacuation.
- Crowd Management and Safety.
- People's Behaviour in Public.
- Venue regulations.
- Reporting procedures.

Documented records of training must include:

- The date of the training session or exercise.
- The duration of the training session or exercise.
- The name of the person giving the instruction or training.
- The names of the trainees or persons receiving the training or instruction.
- The nature of the training or instruction.

Within the code of conduct for stewards it must be made clear that stewards must not leave their place of work without permission, and that they must not consume, or be under the influence of, drink or drugs whilst on duty, and that they must remain calm and courteous at all times.

The stewarding company should provide a detailed 'Statement of Intent'. This document will set out exactly what services the company are providing together with a schedule giving details of all steward positions and staffing levels. All work should be carried out under a written contract signed by both parties before the event; both parties should have copies of the signed contract.

It is particularly important that safety stewards are trained in fire prevention, basic fire fighting, and evacuation techniques. They can also be enrolled to assist with keeping fire lanes clear.

For outdoor events using tents and marquees and all indoor venues, Entertainment Licence conditions will generally dictate the number of stewards to be deployed at entrances and exits to buildings and structures.

Risk assessment techniques and the guidance set out in the Event Safety Guide will clarify correct steward staffing levels for other aspects of the event.

We all have a legal responsibility to look after ourselves first and foremost, and this includes event stewards. It is important to understand that it is illegal for Health and Safety policies and company rules to insist that a steward remains in position whatever the circumstances. If a steward finds himself in a position of personal risk, the steward can, and should, remove his or her self to safety. Such instances may be identified as occasions when he or she is threatened or if his or her safety is at risk in any way.

Risk assessments for stewards should cover specific information and guidance for dealing with violent and aggressive people, lifting people during 'rescue' from in front of the 'pit barrier', noise, and crowd movement. Remember, you can't control a crowd but you can safely and effectively manage one.

The site itself will also need to be risk-assessed for vulnerabilities likely to be a hazard to the public. This may include streams, ditches, lakes and ponds, objects that people could be crushed against in crowded conditions, including gate posts, walls, trees and fences.

A couple of years ago I was involved with an event for over 100,000 people to be held on New Years Eve in a city centre around new dockside developments. My role as Safety Adviser for the event involved risk assessing street furniture, including park benches, bus shelters, kerbstones, lampposts and phone boxes, not to mention water hazards and quayside furniture!

It is important to remember that the majority of health and safety professionals (including licensing officers from the local authority) receive little or no training in crowd safety and crowd management. Therefore it follows, that for events to be properly and safely managed, two safety officials will be required: one a safety adviser, the other a crowd safety manager. Local authority licensing officers also need training in crowd management if they are to understand and carry out their job effectively.

The safety officer and the crowd safety officer will be required to work closely together but it is unlikely that both roles can be carried out by one person as they will probably lack the relevant skills to carry our both roles and will be spreading themselves too thinly to be properly effective in either role.

Crowd safety and management are now developing into a scientific study and computer models are now beginning to be used to calculate flows and density.

A stewarding company carrying out a crowd management role will need to carry out a site survey/risk assessment before they can give a realistic quote for their services. The survey will need to include the following points:

- Staffing levels
- Safe methods of ingress
- Control of localised density
- Control of cultural behaviour
- Safe methods of egress
- Emergency ingress and egress
- Audience profile
- Artist/event profile
- Traffic management
- Public arrival
- Camping and accommodation
- Ground conditions
- Perimeter fencing/walls
- Ticketing/access control systems
- Lighting
- Capacity
- Size of site
- Duration of event
- Disabled facilities
- Possible terrorist attack
- Temporary structures
- Control room
- Briefing areas
- Emergency liaison team (ELT)
- Emergency/contingency planning
- Parking/traffic management arrangements (if the stewarding company is involved in parking/traffic management

These items should all underpin other event documentation.

The stewarding company should be involved in the planning of an event from the outset; they need to give an appropriate and significant input into the overarching event health and safety plan, the site plan, the pit area design, the Major Incident Plan together with all contingency plans.

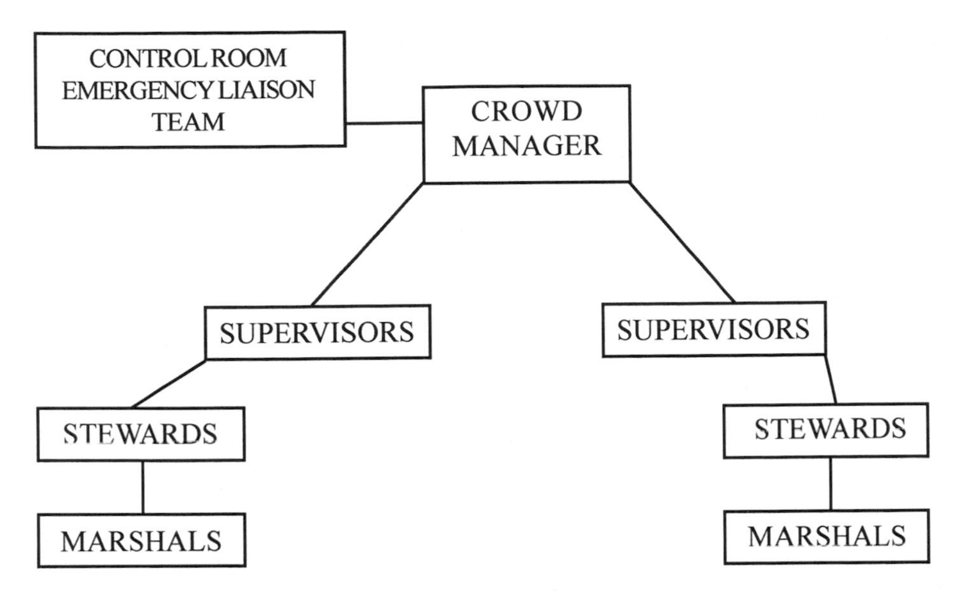

The basic management structure for an event stewarding operation

The stewards at any event are the workforce with the most public interface duties to undertake, the value of their contribution to public relations should not be underestimated. It is of concern that some site/production managers have previously been too arrogant to properly consult with health and safety advisers and crowd managers in producing site plans and drawing up specifications for the pit area.

This type of behaviour, if left unchecked, can only lead to further problems for the future and a lack of credibility within the industry.

Contingency plans need to be fully comprehensive in order to cover every possible eventuality, including a return to normal operation. This can range from cancellation of the event (before the public are admitted or during the event) and 'show stop' procedures, a fire or bomb scare, lost property and children, floods and adverse weather, power failure and leaking diesel tanks. Nowadays, the possibility of terrorism can be added to the list.

I mentioned in chapter 14, Safe Systems of Work, that the contingency plans, should go into detail on the procedures to be followed. It must name the people or groups of people with specific responsibility to effect any required action, as well as providing accurate contact details such mobile phone numbers. All relevant event staff and crew must be provided with copies of the contingency plans in advance of the event so they understand their duties and responsibilities.

If a bomb threat is received at a venue (or premises), it is important that the person who receives the call takes careful and accurate written notes of what is said, listening vigilantly for any accent or any unusual aspect of the caller's voice.

While it is possible that the caller is making a hoax call because she/he holds a grudge against the venue, she/he may be a troublesome punter who has been banned from the venue for instance, or may dislike of the artist performing at the venue or the promoter; any 'code words' that are used must also be carefully noted. It is not for the person receiving the call to make this judgement; the facts must be fully reported to the police.

After receiving the call, immediately dial 1471 (in the UK); this may provide the caller's number. The police should then be called.

It is possible that the threat is made to the police in the first instance, but regardless of who receives the threat the decision to evacuate the premises is normally left to the occupier of the premises or the promoter. The police will not usually make the decision unless they are certain that the call is genuine and not a hoax.

If a decision is made to evacuate the premises, the fire evacuation procedures should be followed. Consideration should be made of the possible location of the explosive device and emergency exits well away from that location should be selected. This procedure must be handled quietly and calmly in order to avoid panic amongst those being evacuated.

All evacuated persons should take their personal property and belongings with them if this does not hinder the evacuation, this will assist the emergency services when searching for devices. Radios and mobile phones should not be used if a bomb is suspected, since radio frequencies could detonate the device.

One contingency plan often overlooked is a method of stopping a show in an emergency: the 'show stop' procedure. I have seen horrific video footage of an event where the crowd collapsed in front of the stage, crushing and killing one girl and injuring many others. On this occasion, the band on stage refused to stop playing and was in fact inciting the crowd to go crazy, stewards in the pit were trying to stop the performance so they could carry out rescues but they were totally ignored. The event safety officer and crowd safety manager must have sovereignty, the authority to stop the show in an emergency without being questioned or involving long debates, systems should be in position to allow an instant 'show stop'.

This is not an isolated incident: several similar incidents have taken place at major events around the world. A simple contingency plan can solve this problem; perhaps it's a role for the tour manager as this is the one person who the band will respect and who will not act without good cause. Regardless of who stops the show, it should be done quickly, simply and without argument. Given the fact that the audience is most likely to listen to, and act upon, information given from the performer, it is important that this aspect of planning is mainstreamed into all event planning and that the artist must be persuaded to cooperate.

A useful and impressive contingency planning tool to which I was recently introduced is the White Light Policy. This could be very useful in some venues, since it works quite simply and can be extremely effective.

If there are any audience problems such as crushing or over-excitement, plain white lights are turned on over the sections of the audience where the problem lies. Plain white static lighting can also turned on over the stage to override the stage lighting effects.

This has a dramatic effect on calming the crowd without stopping the show. As soon as the problem is sorted out, the white lights are turned off and everything returns to… normal!

Remember: A Health and Safety qualification does not make one a crowd safety specialist or crowd manager. The majority of environmental health, police and licensing officers are not trained in, and have little or no understanding or knowledge of, crowd safety or crowd management systems.

With this in mind, perhaps we should question even more the 'competence' of those who are in the position of legislating and licensing our industry.

41 REGULATIONS SPECIFIC FOR CATERERS

Kitchens are crammed full of potential hazards, especial the temporary or makeshift kitchens that tour caterers often have to suffer. In addition to things like noise, electricity, slip and trip hazards, we have hazardous substances, gas, hot objects capable of burning or scalding, knives, cutters that could remove a hand in seconds, and mechanical devices like mixers, etc. Blending all these hazards together with large amounts of liquid creates a wonderful recipe for potential disaster.

Lots of training in Health and Safety as well as all the food hygiene regulations is required for caterers. The law requires all caterers who provide food for anyone other than family and close friends to hold Public Liability (or be covered by Employers' Liability) Insurance and to hold a Certificate in Basic Food Hygiene as a minimum. New catering businesses are required to register with their Local Authority Environmental Health Department, and the Licensing Act 2003 requires catering and food concessions operating between the hours of 11pm and 5am to be licensed. (See chapter two: 'The Licensing Act 2003').

The main legislation for food safety in the UK is the Food Safety Act 1990.

The main regulations under the Act covering catering operations are:

Food Safety (General Food Hygiene) Regulations 1995
Food Premises (Registration) Regulations 1991
Food Safety (Temperature Control) Regulations 1995
Meat Products (Hygiene) Regulations 1994
Dairy Products (Hygiene) Regulations 1995
Egg Products Regulations 1993
Food Safety (Fishery Products) Regulations 1992

To go into detail on these regulations would require another book, so I advise all caterers to get advice from their Local Authority Environmental Health Department, the HSE, the Department of Health and the Food Standards Agency.

There is a requirement for all catering establishments to be registered with their local authority under the Food Premises (Registration) Regulations 1991. This even applies to mobile caterers who must register with the local authority that covers the area where they are based. Registration must be completed 28 days before you start trading and the local authority must also be informed if you change the activity of your food premises; this must also be done within a 28-day period.

Catering areas should always be equipped with a first aid kit, not forgetting the blue plasters, a fire blanket (ideal for chip-pan fires), and a fire extinguisher – foam extinguishers are ideal for fires in deep-fat fryers and most kitchen fires, except electrical fires for which carbon dioxide extinguishers are recommended.

One common complaint about caterers on sites with temporary power supplies regards electrical systems. Caterers always have a terrible habit of overloading socket outlets with the use of multi-way adaptors and extension leads and then wondering why it's 'tripped out' and all gone dead!

Please, caterers: one plug to one socket outlet, no multi-way adaptors, no extension leads and please, please have all your electrical equipment checked, inspected and tested by a competent electrician at the very least once a year; after all, it is a legal requirement!

One question I am often asked by tour caterers is about the facilities they are provided with and the rooms and buildings they are expected to prepare and cook food in, because often these fall a very long way short of the requirements set by the regulations for food safety and hygiene. The question I am asked is: 'should they go ahead and prepare and cook food in premises that do not meet the standard set by the regulations?'

I think most caterers already know the answer to this question but they seem to be in a moral dilemma about what to do when they have the mouths of a lot of hungry crew to feed.

The regulations are very clear that food premises must be suitable and hygienic; they also set out very clearly the basic minimum standards below which you must not drop. You can be prosecuted if the standards are not met and the penalties for non-compliance with the regulations are very high.

My advice is to set out in writing to your client what your minimum requirements are, what you expect to be provided with, and the reason why you can't work with anything less. This should be done before the start of the tour or event and should be one of your contract terms and conditions.

Water

The Event Safety Guide offers guidance on water supplies for events and this advice should be followed. Plumbers also need qualifications to enable them to demonstrate competence. The qualifications for plumbers are as follows:

- A City & Guilds Certificate in plumbing.
- An NVQ Level 2 Certificate in Mechanical Services Engineering. (Plumbing) Level 2 Ref: Q1014057 and/or Level 3 Ref: Q1014058.
- A completed formal Apprenticeship in plumbing.

Further information can be obtained from the Institute of Plumbing.

42 COMMUNICATIONS

Effective communications are a fundamental element within an efficient, successful and professionally presented safety structure. Safety information can be multifunctional and multifaceted; reliant upon the skill and ability of the operator for maximum impact. Communications is a complex area and needs special attention, this chapter only starts to touch the surface. The type and range of this communication can be illustrated by the following list (which is in no way exhaustive):

- 'Post-it' notes
- Labels on a flight cases showing contents, weight and centre of gravity
- Production Schedules
- Site Plans (that should be gridded for accuracy in passing location information)
- Engineering Drawings
- Day Sheets
- Riders
- Itineraries
- Contingency Plans
- Contracts
- Manufacturers' Instructions
- Method Statements
- Risk Assessments
- Health and Safety Policies
- HSE Publications including Regulations and Guidance
- Approved Codes of Practice
- Signs

In order to receive and transmit information, most of us use quite sophisticated communication systems on a daily basis. Typically, we use the Internet, PA systems, phones and fax. Accurate and professional use of this equipment in an event situation is essential as they are key tools for both the event organiser and production manager; they must never be underestimated as a safety tool. Both written and verbal communications remain at the heart of human communications systems, but we must learn to develop and embrace new technology that enhances and maximises opportunities whilst learning to use equipment correctly and building on existing communication skills.

Radios are another instrument in common use at events. However, many of us are uninformed about the correct way to use these radio systems. I have overheard all sorts of rubbish being talked over these airways together with a lot of unintelligible dialogue. The problem with incorrect usage of an essential communication system is that, had there been an emergency, it may have resulted in fatal consequences simply because the message would have been blocked from reaching its destination.

If a communications system is to be effective, it needs careful planning. A communications manager may be required at a large event to coordinate the whole operation and all messages will need to be recorded at a central control position.

Anyone who has worked in an event radio control room will know just how important a well-run control room is and how stressful this work can be.

An event control room is often a multi-agency operation reflecting the joint working practices and partnerships that underpin effective management of large events. These partnerships will typically include such organisations as the police, the fire service, the ambulance service, and the local authority, all working in conjunction and close cooperation with the event organisers, traffic manager, safety officer, production manager, site manager and crowd safety officer. This same group may also form the core of the emergency liaison team.

Field telephones, C.B. radio and amateur radio systems will often be considered a probable means of effective communication for large outdoor events. This form of communication is most effective when used in conjunction with a dedicated radiotelephone system.

Where information has to be communicated to large crowds and no PA system exists, the use of loudspeakers has proved to be an efficient and effective alternative system. Large video screens or video walls are a particularly effective manner of disseminating information to large crowds and should, therefore, be considered as a useful element within the overall crowd management system.

Where PA systems and video screens/walls are to be used for emergency announcements, procedures should be put in place for dealing with such announcements. Stage managers, sound crew, video crew, the artist and management team, together with announcers, need to be fully briefed of these procedures in advance. Similarly, a simple procedure should be in place for stopping a show or performance in an emergency.

Prior to a large outdoor event, tests should be carried out to see if suitable signal strength exists for mobile phones. Service providers might consider installing temporary masts and booster systems to low-signal areas for large events. Even in areas where a good signal exists, it still may not be possible to make and receive calls at certain peak times due to the cells being overload by the huge extra numbers of users – New Year's Eve 2000 springs to mind here – it is for this reason that mobile phones should not be relied upon as a primary means of communication.

It is important that PA companies and those supplying or operating radio microphones only use systems licensed by JFMG Ltd., who are contracted by the Radio communications Agency, an executive agency of the D.T.I., to manage and licence radio systems used in programme-making and special events. Unlicensed radios may operate on frequencies that interfere with the emergency services and you may get interference and messages from the local radio mini-cab service being transmitted over the PA. This obviously means yet more work for production managers checking on radio licences.

Correct Use of Radios

Operating a radiotelephone is extremely simple; over a quarter of a million people in the UK alone use them every day. Keeping it simple but effective means following a few simple rules.

The subsequent information is intended as a guide to best practice when communicating by radiotelephone. The purpose is to foster a better understanding of the need for practical rules so that users can work to an improved capacity. Those working in areas where high sound pressure levels exist, such as on stage, will require headsets with their radios for audio clarity.

To make a call

1. Make certain that your set is turned ON; many users forget to do this, listen to Control and set the volume to a comfortable level.

2. If you have a multi-channel set, make sure you are tuned to the correct channel.

3. Important: *listen* before transmitting. If you are not sure whether the channel is in use, briefly ask.

4. Never interrupt another station already working the channel except in case of emergency.

5. Do not all speak at once.

6. When you are certain that the channel is clear, press the PRESS TO TRANSMIT button and speak slowly, clearly and softly across the microphone, not directly into it, holding the microphone about 1-2 inches away from your mouth. Do not shout or whisper. The radiotelephone has all the amplification it needs and speaking above a certain level distorts the signal. Let the radiotelephone do the work. Normal speech speed is about 40-60 words per minute. Pronounce words *disctinctly* and *slowly.* Our speech tends to speed up when we are nervous and stressed, so take a deep breath before speaking (this applies to telephones as well as radios).

7. At the end of your transmission release the PRESS TO TRANSMIT button to listen to the reply.

The use of obscene language, swearing, and the passing of gambling information is expressly forbidden under Radio Telephone Licensing Regulations. The police or Home Office may be monitoring your calls, and in some cases event stewards may be operating on the same channels as the police with whom close liaison is essential.

The Standard International Phonetic Alphabet is recommended when it is necessary to spell out individual words. It is mainly of advantage when operating in difficult conditions of communication and is always given as 'A for Alpha', B for Bravo', etc. Never as 'A as in Alpha', B as in Bravo', etc.

Numerals should be pronounced as shown in the table opposite.

Strict adherence to procedure is necessary to ensure *accuracy and speed* in passing messages and to eliminate mistakes and repetitions. In emergencies, think before transmitting and put facts slowly and clearly to Control; your professional input is vital.

A	Alpha
B	Bravo
C	Charlie
D	Delta
E	Echo
F	Foxtrot
G	Golf
H	Hotel
I	India
J	Juliet
K	Kilo
L	Lima
M	Mike
N	November
O	Oscar
P	Papa
Q	Quebec
R	Romeo
S	Sierra
T	Tango
U	Uniform
V	Victor
W	Whiskey
X	X-Ray
Y	Yankee
Z	Zulu

1	Wun
2	Two
3	Three
4	Four
5	Fife
6	Six
7	Sev-en
8	Ate
9	Niner
0	Zero

Emotional tension can speed up the word rate and reduce intelligibility to the point where Control and you are just not communicating. *Slow down and speak intelligibly*. Use code names, numbers and abbreviations. Avoid jargon which may not be understood.

Do not use a radiotelephone for idle conversations, since transmitting uses up batteries more quickly than receiving. You will be able to 'monitor' your radio for several hours, but you have very limited transmission time, and a flat battery is of no use to you when you have an urgent message to transmit. Chatting unnecessarily on your radio is unprofessional and sets a poor example to colleagues and members of the public. If you spend your work time chatting you will be out of contact with Control, blocking the channel, and be oblivious to any situation that may require your urgent attention. It is no exaggeration to state here that lives may be at risk.

Call Procedure

After first making sure the channel is clear, call once using the full call signs and always giving your own call sign after that of the station you are calling.

Example:

If Alpha 1 wishes to call Alpha Control:

Alpha Control, Alpha Control, Alpha Wun Over.

Messages

Be as brief as possible, using abbreviations and codes, ending each time with 'over'.

Example:

If Alpha 1 wishes to call Alpha Control:

Alpha Control, Alpha Wun, Job completed. Any message? Over.

Acknowledgements

Every message must be acknowledged, otherwise the calling station may think the exchange is incomplete and may try to hold the channel clear of other users.

Closing

1. At the end of a series of messages both stations should finish with the words 'standing by', indicating that no reply is expected and that they are open to receive messages from other stations.

2. The words 'over and out' are only used when a station is closing down and going totally off the air and no further messages are to be transmitted or received.

Control	Alpha Wun, Alpha Wun, Alpha Control. Over.
Alpha One	Alpha Control, Alpha Wun, Wait Wun. Over.
Alpha One	Alpha Wun, Go ahead. Over.
Control	Alpha Wun, What is your E.T.A. at Glastonbury? Over.
Alpha One	1600. Over.
Control	Roger, Alpha Control Standing By.
Alpha One	Alpha Wun Standing By.

Example of procedure.

E.T.A.	Estimated time of arrival
E.T.D.	Estimated time of departure
R.T.B.	Return to base
Say Again	Return to base
Roger	Message understood
Over	Used at the completion of an exchange of messages
Wait	Indicates that you are unable to reply immediately
Go ahead	Normally used after a 'wait' period

Example of abbreviations.

In an emergency situation, it may be necessary to interrupt other users on the same channel in order to convey an emergency message. You must interrupt existing conversations by using the words "PRIORITY, PRIORITY" this makes it clear to all other users you have an emergency message. Hearing these words all other users must cease conversation immediately in order to allow the emergency message to be transmitted. Radio silence must be maintained until the emergency situation has been fully dealt with and control gives the 'all clear' to resume normal use.

At the end of your shift, turn off your set and put the battery on charge ready for the next shift. When an extra long shift is anticipated, make sure that you carry a spare battery. At the end of your duty return all radios and equipment and see that is checked in properly, you may be held responsible for any loss!

Operators of base stations will need further training beyond the scope of this book.

After any event it is vital that a detailed debriefing takes place amongst all departments so that any lessons can be learnt and mistakes rectified prior to future events. The complexity of the debriefing session will depend on the size and scale of the event.

98% of all work-related accidents are due to human error!

43 WORK-RELATED STRESS

What is Stress?

Stress is the adverse reaction people have to excessive pressure or other types of demand placed on them. Things at work or things outside work, often both, can cause stress. Stress is not an illness but it can lead to serious health problems such as heart disease, diabetes, back pain, being overweight and other illnesses. Stress can also bring on psychological problems like anxiety, depression or nervous breakdown.

The symptoms of stress include changes in a person's mood or behaviour, deteriorating relationships, irritability, indecisiveness, absenteeism and reduced performance at work. It is common for victims of stress to turn to smoking, drinking and/or drugs, or they may complain about their health more frequently.

Being under a little pressure is not always a bad thing, but stress is very bad for you. We are all vulnerable to stress, but it is safe to assume that most of us are fairly hardy and can stand some pressure. However, employers should be on the lookout for anyone who is particularly vulnerable because of work pressures or how work is organised. Steps must be taken to reduce the pressure on these people.

As well as pressure of work, other factors that can induce stress include problems such as finances, moving house, family, bereavement, and relationships; the victims of bullying, harassment, violence, aggression and prejudice are also prone to stress.

Dealing with Stress

Employers do not have a responsibility to deal with stress problems that originate from outside the workplace, but it helps considerably and is in the employer's interest if they are sympathetic, understanding, and try to offer some help. Employers do have a duty to prevent employees from becoming ill at work and work stress causes ill health. Employers must take stress very seriously.

The cost of reducing work-related stress is very low and very cost-effective, far less than a compensation claim from an employee who is suffering ill health due to work- related stress. Work-related stress can usually be prevented in the first instance and removed or reduced where it already exists with good management systems and techniques. Allowing sufficient time for the planning, pre-production and production of shows and tours is the first major step to reducing stress and raising health and safety standards. Good management will also ensure that touring and event personnel have accommodation and catering of a reasonable standard.

It's probably no surprise to discover that risk assessments are again on the agenda. This time we are looking for pressures at work that can cause high and long-lasting levels of stress. One must then assess who is at risk, and then decide if the existing controls are suitable and sufficient or if more controls need to be introduced and implemented. As with all risk assessments, the final stage is recording and then regular monitoring.

Employees must be involved at all stages of the process. The first step that can be taken is to talk to staff to express what problems there may be. For instance, they may be disillusioned with their work, or they may be over-trained for the job they are required to do.

Talking and listening to staff is the first important step, so try asking them what the three best aspects and the three worst aspects of their job are. Use the information collected to help identify common and persistent work pressure problems and whom this pressure may affect.

Remember to respect the confidentiality of staff, to tell staff what you are doing and what you intend to do with the information collected, to involve them in subsequent decisions, and to involve safety representatives, if you have them in your organisation.

Employees may be reluctant to talk about stress or admit they are stressed, since they may see stress as a weakness. Managers and employers can help by leading by example, by making it easy to talk about stress and reassuring staff that all information will be treated in confidence.

Stress can be reduced or removed in various ways. Management must first demonstrate that they take stress seriously and are understanding to those who admit to being under pressure. Employers and managers must have an open and understanding attitude to what people say to them about work pressure and must be able to identify the signs of stress in staff.

Ensuring that staff have the training, skills and resources (including time) is most important. Staff must know what they must do, are confident they can do it and get credit for the work they do; involve staff in decision-making as well. If any member of staff thinks he/she is already suffering from stress, get them to seek medical advice. If an employee is off work sick, then keep in contact with them and with their doctor. When they can return to work they may have to do a different job, work shorter hours or only do part of their old job before they can return to their old job full time. Do not tolerate bullying or harassment of any kind in the workplace and ensure that people are treated fairly and consistently. If stress can't be removed or reduced, consider moving an employee to a different job.

If possible, allow staff to influence how their jobs are done, vary the working conditions and flexibility; this will increase interest and sense of ownership. Ensure good levels of communication between staff and management and don't be afraid to listen. Try to address the source(s) of pressure rather than dismissing a member of staff, which could lead to an unfair dismissal case.

Problems that can lead to work-related stress:

- Lack of communication and consultation.
- A culture of blame when things go wrong, denial and potential problems
- An expectation that people will work excessively long hours or take work home.
- Too much to do, too little time.
- Too much or too little training for the job.
- Boring or repetitive work, too little to do.
- The working environment.
- Lack of control over work activities.
- Staff feeling that the job requires them to behave in conflicting ways at the same time.
- Confusion about how everyone fits in.
- Poor relationships with others.
- Bullying, sexual or racial harassment.
- Lack of support from managers and co-workers.
- Unable to balance the demands of work and life outside work.
- Uncertainty about what is happening.
- Fears about job security.

In most cases, complex stress management courses, Employee Assistance Programmes and counselling are not necessary but can be useful as part of the 'bigger plan' to tackle work-related stress.

44 CONSTRUCTION (DESIGN AND MANAGEMENT) REGULATIONS 1994

This regulation was only really relevant to staging companies and companies building structures. I say 'was' because the HSE has stated that in the majority of cases these regulations do not apply to our industry. This is supported by a statement in the Event Safety Guide, but this situation may reverse in the future and we may find outdoor sites classed as building sites and construction site regulations imposed on us.

It is very interesting to note that just because a particular regulation does not apply, it does not mean that 'we don't need to do anything'. The big umbrella of the Health and Safety at Work Act still applies, as do the Health and Safety Management Regulations. HASAWA and the Management Regulations actually require you to do almost everything that is required under the CDM Regulations, so in some ways the HSE decision is meaningless.

The CDM Regulations were brought in to make clients and the contractors they engaged more aware of their responsibilities towards Health and Safety in the construction industry. It also brought in a few new responsibilities and created a new position, that of a Planning Supervisor. It is obvious that making the client (promoters/artists) more aware of their responsibilities should do a lot of good for the likes of production managers, contractors and employees.

The CDM Regulations are in fact just a new set of regulations that just 'tighten up' other regulations (most notably the Health and Safety Management Regs.) that we should be complying with anyway; there is very little in the regulations that is actually new; CDM puts roles and responsibilities to existing regulations. It's nothing really new and nothing to get worked up about – if you're following good health and safety practice from the outset it will have very little effect on you.

Other than the new responsibilities and the need for a planning supervisor, the CDM Regulations are a good example of how Health and Safety should be managed, and what the law expects. However, it does produce a lot of extra paperwork in addition to the requirement for suppliers to submit risk assessments, method statements, insurance, full details of staff training and

Health and Safety control systems for the job in question to the client at the same time as a quote. As we have seen, this is as it should be so the client can then assess this information and decide who the most suitable and 'competent' contractors to appoint are.

It's quite easy to transpose the CDM Regulations to our industry. We still have the client (promoter or artist/artist management), the planning supervisor becomes the event safety officer or adviser, the designer becomes the production manager, the staging company could be principal contractor, and we still have other contractors and self-employed (PA, lighting and crew). A tour or major event should still have a Safety File as described in the CDM Regulations to comply with the HASAWA and the Health and Safety Management Regulations, even though CDM does not apply.

This safety file should be held by the safety officer, but all the service companies involved (and the self-employed) should contribute towards it.

Similarly with most of the regulations under the HASAWA, if the HSE or local authority enforcement officers are really trying to enforce and prosecute, it is unlikely that they would try to do so under the CDM Regulations (even they did apply). It is more likely that they would try to do so under the HASAWA as it carries far greater penalties and is easier to enforce.

It has become quite obvious to the author that the skeletal framework of the CDM Regulations is to become a standard management framework for the HSE to impress on other industries. I'm sure that those readers who are really getting a grasp of all this will realise that this gives us the opportunity to be one jump ahead and get our own management frameworks in place. Like the Lifting Operations and Lifting Equipment Regulations, I think CDM is what our industry really needs; perhaps not in its present format (as applied to construction), but with changes to suit our industry and with a change of name to, say, 'EDM' (Event Design and Management Regulations) and, hopefully, a reduction on the amount of paperwork required. I believe it would go a long way in forcing out the cowboys and rogue promoters by self-regulating. Isn't that just what we want?

All that CDM does is put roles and responsibilities to existing regulations – used properly, it can make life easier for us!

45 YOUNG PEOPLE AT WORK

The Health and Safety (Young Persons) Regulations 1997 make certain requirements on employers who employ these individuals. Young workers are particularly at risk because of their possible lack of awareness of existing or potential risks, immaturity and inexperience. Children under 13 years old are generally banned from any form of employment. Children between 13 and the minimum school leaving age (16 years old) are prohibited from being employed in industrial undertakings such as factories, construction sites, etc, except when on work experience schemes approved by the Local Education Authority. The author suspects that children between 13 and 16 are also prohibited from working in the area of live production because of the number of hazards and the high levels of associated risk.

Employers are required to:

- Assess risks to young people, under 18 years old, *before* they start work.
- Take into account their inexperience, lack of awareness of existing or potential risks and immaturity.
- Address specific factors in the risk assessment.
- Provide information to parents of school-aged children about the risk and the control measures introduced.
- Take account of risk assessment in determining whether the young person should be prohibited from certain work activities, except where they are over the minimum school leaving age and it is necessary for their training, and
- Where risks are reduced so far as is reasonable practicable, and
- Where proper supervision is provided by a competent person.

Young persons must be protected from:

1. Work beyond his/her physical/psychological capacity.
2. Exposure to harmful agents (toxic, carcinogenic).
3. Risks that cannot be recognised or avoided by young persons due to their insufficient attention to safety or lack of experience or training.
4. Risk to health from extreme heat or cold, noise or vibration.

5. Harmful exposure to radiation.

New proposals were issued by the DTI that became law on 6th April 2003. Under the new law young people (between the minimum school leaving age and 18) will be prohibited from working between the hours of midnight and 4am unless the work involves:

- Employment in healthcare establishments – such as hospitals.
- Employment in connection with cultural, artistic, sporting or advertising activities – such as the performing arts.
- Work in shipping or fisheries, or
- The armed forces.

In addition, the new law has introduced a ban on people under 18 years of age working between 10pm and 7am unless the work involves:

- Work in the retail, hotel catering and bakery industries – including bars, restaurants and nightclubs.
- Agriculture, and
- Postal services and newspaper deliveries.

Furthermore, a young person can only be employed during these hours if there is:

- A need for continuity of service or production or to respond to a surge in demand for a service.
- No adult worker available to perform the work, or
- Performing the work would not affect the young worker's education or training.

The amount of time that a person under 18 years of age can be required to work will be further reduced under the new law to a maximum of 8 hours per day and 40 hours per week. However, a young person will be exempt from the new law if there is:

- A need for continuity of service or production or to respond to a surge in demand for a service.
- No adult worker available to perform the work, or
- Performing the work would not affect the young workers education or training.

Further regulations cover young persons under the school leaving age, but this is beyond the scope of this book. Further information may be obtained from your local Department of Works and Pensions office if you have workers or intend to employ workers in this age group. The hours they may work and the work they are allowed to do is very limited.

46 OFFICE SAFETY

As we have already discussed, to be effective our health and safety 'culture' must be established at the top of a company's management structure and filter down to the levels below. This 'culture' must also be firmly supported by policy and management systems. The office or operations base is no exception and these areas must reflect the culture and company policy from the start. This can become a dilemma to an organisation that may be establishing its safety management systems and policy for the first time. If you are unsure of where to start, carry out a simple risk assessment to ascertain where the greatest hazards and risks lie. Now start there.

With the exemption of CDM Regulations, Fall Arrest, Lift Trucks and Plant, Special Effects and, Lifting Operations and Lifting Equipment Regulations and perhaps the LPG Regulations, most of the regulations we have looked at also apply to our offices. In addition, if computers are used in the office than you may need to do Work Station Assessments as required by the Display Screen Equipment Regulations 1992 and then 'action' any controls as required.

The Workplace Health, Safety and Welfare Regulations 1992 apply very much to offices. These regulations also require the room temperature to be at least 16°C (after the first hour) where people work sitting down (e.g. offices) and for a thermometer to be available to persons at work to enable temperatures to be measured throughout the workplace. Also, consider smoking as a serious health hazard. In several countries smoking is now banned in all work places.

Owners and managers of small, low-risk office premises where people are employed should be aware of the requirement to register using form OSR1 that can be obtained from your enforcing authority (generally your local Environmental Health Department). This form asks only for very basic information such as your correct address and how many people you employ; it only needs to be filled in once.

The point to remember is that health and safety starts at the top and works down. The 'top' for most employees is the boss or MD who is usually office-based, so your health and safety cultural change will usually have to start here and in the warehouse, workshop and yard if you have them.

47 THE HEALTH AND SAFETY (DISPLAY SCREEN EQUIPMENT) REGULATIONS 1992

Prior to 1992 there was no existing legislation in this area, so these regulations impose new, long-overdue duties on employers. All employers will now have to take appropriate measures to prevent employees from suffering repetitive strain injuries (RSI), aches and pains in the hands, arms and wrists – collectively known as 'work-related upper limb disorders' (WRULDs) – and damage to eyesight as a result of long exposure to display screen equipment. The symptoms of WRULDs can include temporary body fatigue and soreness to chronic soft tissue disorders such as Carpal Tunnel Syndrome. Research has shown that DSE has no effect on pregnant women and that no risks are present from radiation, furthermore, epileptics and even those who suffer from photosensitive epilepsy find they can safely use DSE.

These Regulations cover all persons who use display screen equipment as a significant part of their everyday work. The definition of 'display screen equipment' covers computers and microfiche as well as television and film pictures, including CCTV operators and film and video editing.

There are different standards of protection for 'users' and 'operators'. Broadly speaking, the 'operator' is self-employed and the 'user' is an employee – Regulation 1(1)(b) and 1(1)(d). The duty to users covers not only duties to employees but also employees of other employers, e.g. agency staff.

Generally speaking, users are:

- Those who use display screen equipment more or less continually during the working day or for continuous spells of an hour at a time on a daily basis, and
- Those who have to transfer information quickly to and from the screen.

Home-workers and those who work away from base (often with a laptop computer) should also be classed as users, but the self-employed are defined as operators and are only covered by certain parts of the law.

The Employer is Required:

- To carry out Health and Safety assessments of the workstations in his undertaking used by operators or users [Regulation 2].
- To reduce the risks revealed by the assessments to the "lowest extent reasonably practicable" [Regulation 2 (3)].
- To ensure that all workstations put into service after 1st January 1993 meet the requirements set out in the Schedule to the Regulations.
- To ensure that all workstations in use prior to that date comply with the requirements *now* [Regulation 3].
- To plan periodic interruptions in the work of 'users' of the display screen equipment [Regulation 4].
- To provide eyesight tests for employees [Regulation 5] and to provide, where necessary, the appropriate corrective appliance, for example, glasses or contact lenses [Regulation 5(5)].
- To provide adequate training on the health and safety aspects of the workstation, and to ensure that adequate training is also given whenever the organisation of the workstation is substantially modified [Regulation 6].
- To provide information on health and safety matters to all users and operators [Regulation 7].
- To reduce the risk of fatigue or WRULDs, the use of DSE should be planned so that DSE work can be broken up with other tasks; a mixture of DSE and non-DSE work will vary visual, physical and mental demands and will reduce risk to the user or operator.

Users should be reminded to stretch, blink and change position frequently. The use of a mouse can tend to concentrate activity on one arm or finger, and intense use of a mouse can lead to aches in the fingers, hands, wrists, arms or shoulders.

Operators and users should be encouraged to take frequent breaks in addition to imposed breaks; short, frequent breaks are preferable to longer, infrequent breaks. Breaks should be taken before the user becomes tired rather than as a method of recovery.

Employers are required to provide and pay for eyesight tests for all employees using DSE and those who are about to become users. Eyesight tests need only be provided if the user requests it and they do not have to be provided for self-employed workers. Users are also entitled to further tests at regular intervals after the first test, and in-between if they have vision difficulties that could be attributed to DSE.

If the test shows the user requires glasses for DSE work, then the employer is obliged to pay for a basic pair of frames and lenses. The user can 'top up' (from their own pocket) the employer's contribution to pay for a more expensive pair if the employee so wishes.

Users and operators of laptop and other portable computers need to take special care, because they often have small keyboards and screens that make them less comfortable to use for long periods of time.

Training in the safe use of DSE should include:

- Information about the risks from DSE.
- The importance of good posture and how to adjust furniture to avoid risks.
- The need to organise activity changes and screen breaks, to change position, to stretch and blink.
- How to organise the workstation to avoid stretching movements.
- How to avoid reflections and glare on the screen, including how to adjust and clean the screen.
- The need to cooperate in the risk assessment process by completing questionnaires and checklists.
- Details of the person within the organisation responsible for implementing the regulations.
- The assessment is usually made through the use of a simple checklist by a 'competent' person who is familiar with the regulations and is able to:
    - Assess risks and the kind of display screen work being carried out.
    - Draw upon additional sources of information as required.
    - Draw a valid and reliable conclusion from the assessment.
    - Record the assessment in writing and communicate the findings to those who need to take appropriate action.
    - Recognise their limitations so that, if necessary, further expertise can be called upon.

The checklist will consider:

- If the workstation is correctly laid out and of suitable size.
- The frequency and duration of the work.
- The lighting in the area of the workstation and that appropriate

contrast between the screen and background environment exists.

- If the screen is adjustable for brightness and contrast.
- If the software is user-friendly.
- If the displayed image is stable with no flickering.
- The characters on the screen are of adequate size, clearly formed and defined.
- The screen and keyboard can be adjusted to suit the needs of each user or operator.
- There is no glare or reflection on the screen that could cause discomfort to the user or operator. Light sources such as windows, skylights, brightly coloured fixtures and walls can all cause problems that need to be eliminated.
- That there is space in front of the keyboard large enough for the user or operator to support hands and arms.
- That the keyboard has a matt surface and does not cause any reflective glare.
- That there is sufficient space beneath the workstation for legs and feet to move in comfort; those employees who request a foot-rest should be provided with one free of charge.

Finally, the user's or operator's chair must be stable, adjustable, allow freedom of movement, and be positioned so that uncomfortable head and eye movements can be avoided. The chair should be adjustable in height, while the seat-back should be adjustable in both height and tilt.

48 MEASURING PERFORMANCE

Companies that achieve high standards in health and safety do so by setting performance standards and continually assessing the extent to which those standards are being achieved. Monitoring your company's health and safety performance enables you to:

- Assess the effectiveness of the policies and procedures you have in place.
- Identify areas where additional safeguards are necessary.
- Fulfil your legal record keeping requirements.
- Demonstrate your ongoing commitment to maintaining a safe and healthy workplace.
- Of course records of monitoring activities could also prove extremely valuable in the event of legal action or an insurance claim.

Active and Reactive Monitoring

Effective monitoring systems incorporate aspects of both active and reactive monitoring. *Active* monitoring measures performance against pre-determined standards to ensure that systems and procedures are being properly implemented and that standards are being achieved. Active monitoring activities are concerned with prevention rather than cure, in that their aim is to identify areas of poor performance before an incident occurs. *Reactive* monitoring is concerned with investigation of incidents to find out exactly what happened, why it happened, and how it can be stopped from happening again.

Active Monitoring

In your active monitoring activities consider the following:

- Are performance standards being achieved?
- Are current control measures being properly implemented?
- Is all Personal Protective Equipment being properly used?
- Do all employees know and obey the company safety rules?

There are various methods of discovering the answers to these questions:

Workplace inspections

Regular random workplace inspections will show whether work equipment is being used properly and if safe working practices are being observed, as well as keeping staff 'on their toes'.

Questionnaires

Questionnaires are a useful tool for discovering employees' views on health and safety performance and can also be used as a test of how effectively you are getting the safety message across to them.

Random Observation

Observation of working practices should be done from a discreet vantage point. Employees are likely to obey rules they might otherwise disregard if they know they are being watched.

Environmental Checks

Regular checks on such things as noise levels, air quality, the state of the washroom floor, etc. will soon show if performance standards are being achieved.

It is important to concentrate your active monitoring activities in the areas of highest risk. Consider where the most serious harm could be done and pay particular attention to those areas/processes.

The information you collect from your active monitoring activities should form the basis of future performance objectives.

Reactive Monitoring

No matter how comprehensive your accident prevention programme is it is almost inevitable that accidents will happen. People take shortcuts, forget to use protective equipment and fail to pay attention to what they are doing. Reactive monitoring is concerned with analysing incidents as they happen, and should aim to discover:

- Exactly what happened
- Why it happened, and
- How it can be prevented from happening again.
- While this may seem like locking the stable door after the horse has bolted, it must be recognised that accidents do happen and we should endeavour to learn from our mistakes. Indeed many lessons can be

learned from experience, and things that seem obvious with the benefit of hindsight are often overlooked.

- Your reactive monitoring activities should centre around the study of:
  - Accident reports
  - Attendance sheets
  - Inspection records

Accident Reports

Under the Reporting of Injuries, Diseases and Dangerous Occurrences Regulations 1995 (RIDDOR 95) you must keep records of reportable incidents. As a matter of good practice though, you should keep records of all accidents, no matter how minor, as these will enable you to assess more accurately your company's health and safety performance.

For reactive monitoring to be effective there must be an efficient record keeping system in place. Records must be comprehensive and consistent so that they can be compared. You should therefore develop and maintain systems for recording details of:

- Accidents and near misses
- Hazards (workplace inspection/risk assessment records)
- Attendance.

Provision should be made for a system for recording details of hazards, accidents and workplace inspections.

Recording Accidents

Details of any accidents should be recorded on the correct form as soon as possible after the event. The essential details to be recorded are:

- The time, date and location of the accident.
- Details of any injured person(s) including name, job title and department.
- Details of any injuries sustained.
- What happened, including the chain of events leading up to the incident.
- Details of any damage to property, equipment, etc.
- Details of witnesses and witness accounts of the incident.
- The cause(s) of the incident, including direct and indirect contributory factors.

- Action taken to prevent recurrence.
- Further action needed to prevent recurrence.
- In some cases, it may be appropriate to attach diagrams or photographs to the report.

Near Misses

It is important that near-miss incidents are treated in the same way as accidents in terms of reporting and investigation. A near miss is defined as an unplanned and unforeseeable event that could have resulted, but did not result, in human injury, property damage or other form of loss. Not all incidents will merit immediate thorough investigation, but if the potential outcome of the near miss was a serious accident then the incident should be investigated as a matter of priority.

Attendance

Absence from work due to sickness is one indicator of health and safety performance which is often overlooked by companies in their monitoring activities. There may be numerous reasons for employee absenteeism but patterns of absence can indicate areas where the organisation could improve the situation. In particular, you should look for patterns such as:

- Higher than average absenteeism rates in a certain department.
- Employees regularly having time off for the same medical complaint.

It is therefore important to elicit details of employee illness, and a self-certification sickness form is the best way to obtain this information.

Analysing the Information Collected

An effective monitoring programme should highlight those areas in which your company is under-achieving in the area of health and safety. In some cases the reasons for this failure may be immediately apparent, in other cases further investigation may be necessary in order to identify the underlying causes. Data analysis should enable you to identify the appropriate course of action. This is likely to include one or more of the following:

1. Improve controls, e.g. introduce new systems of work, install machine guards, issue new personal protective equipment.
2. Investigate further, e.g. conduct a fresh risk assessment, establish a safety committee.

3. Train staff – establish a safety training programme, conduct safety seminars.
4. Conduct a job evaluation study to identify the skills, qualifications and degrees of competence necessary to perform different tasks safely.
5. Improve environmental controls, e.g. reduce noise levels, increase lighting.
6. Increase employee involvement in safety policy formulation.
7. Discipline employees who breach safety rules.

49 CONCLUSION

A few key points should now be obvious. Health and Safety regulations change rapidly. Every time I think I am getting near finishing this book a new change in regulations appears – not always new regulations – it's generally changes and updates to existing regulations to meet European Directives.

Health and Safety is not just about getting the crew to wear safety boots, hard hats and fall arrest equipment, or even about testing odd items of equipment on a periodic basis. It is about putting in place a structured and visible management system and controls based on policy, legal requirements and risk assessment and, having put that system in place, monitoring, assessing and auditing to measure performance, rather as you would with a financial management system.

There is a legal requirement to have such a system in place, but the authorities are very lax when it comes to enforcement, as, unlike financial management systems that are regularly checked as they bring in revenue to the Government, there is very little revenue from Health and Safety enforcement at the moment. However, things are going to change very soon and the three new offences on the statute books, including Cooperate Killing, are the start. Directors and those with a 'controlling mind' in a business operation had better look out as the government is introducing a 44-point action plan that includes:

1. A Code of Practice
2. Compulsory Training
3. Suspension of managers without pay
4. Fines and imprisonment
5. Fines linked to company turnover
6. Suspended sentences pending remedial action.

It is not possible to have a functional safety operating system based upon anything other than this management system. Not only that, but it has to be in place throughout the whole business and not just in certain areas such as the yard or on site. It has to come from the top down; it's a board-room responsibility and decision. The new voluntary code of practice for directors shows the way forward here.

In a way, Health and Safety management is a form of insurance. You can insure against civil claims, against damage, loss or fire, but you can't insure against carrying out a criminal act and a breach of Health and Safety law is a criminal act that carries high penalties. The way to protect yourself and your business is to install a Health and Safety management system. Isn't it stupid to insure against a remote risk but not to insure against a very possible risk, such as when you are involved in dangerous work operations?

The cost of putting such a system in place is very low when weighed against the cost of an accident. There is an initial cost, but very low running costs after the first outlay.

Anyone involved in putting in place such a system must have the full support, cooperation, respect and assistance from management together with access to certain information about the business to enable them to carry out their duties. This can only come about when management and staff have been fully informed and consulted on what is going to take place.

Unfortunately, it may take a few more deaths before the HSE starts to get really interested in our activities. No doubt a few serious prosecutions will follow these deaths, and perhaps then we shall see some real action from hire companies, artists, production companies and promoters.

All we can do is try to get the message across to those who haven't got their act together before any of us ends up in a Coroner's Court or a prison with somebody's blood on our hands!

A number of people have talked in the past about a Code of Practice for our industry. It's clear to me that health and safety legislation, standards and regulations are our code of practice ready-made for us.

Finally, by the time we start to get a grip with basic Health and Safety, there will be a new set of Health and Safety regulations – you can be sure that Environmental Health and Safety will be a major issue very soon. This will involve taking into consideration the effect and impact of our work on the health and safety of the environment and with this will come new environmental health and safety policies, risk assessments and control systems, etc. The Environmental Agency has already started work on this and some regulations are already legislation. For example, the Control of Pollution (Oil Storage) (England) Regulations now require all oil (including diesel) containers of over 200 litres to be bundled (double-skinned or on drip-trays large enough to hold the full capacity of the container) and for 'spillage' kits to be carried in vehicles transporting drums of diesel.

A1 DEFINITIONS AND ABBREVIATIONS

Throughout the HASAWA and the various regulations it covers a few terms keep cropping up. They are: *reasonable and practicable*; *so far as is practicable*; and *best practicable means*.

Although none of these expressions is defined in the HASAWA, they have acquired meanings through many interpretations by the courts and it is the courts which, in the final analysis, decide their application in particular cases.

To carry out a duty *so far as is reasonably practicable* means that the degree of risk in a particular activity or environment can be balanced against the time, trouble, cost and physical difficulty of taking measures to avoid the risk. If these are so disproportionate to the risk that it would be unreasonable for the persons concerned to have to incur them to prevent it, they are not obliged to do so.

The greater the risk, the more you are reasonably expected to go to very substantial expense, trouble and invention to reduce it. But if the consequences and the extent of the risk are small, insistence on great expense would not be considered reasonable.

It is important to remember that the judgement is an objective one and the size or financial position of the employer is immaterial.

So far as is practicable, without the qualifying word 'reasonable', implies a stricter standard. The term generally embraces whatever is technically possible in the light of current knowledge, which the person concerned had or ought to have had at the time. The cost, time and trouble involved are not to be taken into account.

The meaning of *best practicable means* can vary depending on its context and ultimately it is for the courts to decide. Where the law prescribes that 'best practicable means' should be employed, it is usual for the regulating authority to indicate its view of what is practicable in notes or even agreements with particular firms or industries. Both these notes or agreements and the views likely to be taken by a court will be influenced by considerations of cost and technical practicability, but the view generally adopted by HSE inspectors is that an element of reasonableness is involved in considering whether the best practicable means had been used in a particular situation.

The terms 'competent' and 'competency' appear regularly in Health & Safety regulations and Health & Safety Law, but (apart from in the Electricity at Work Regulations and the Work at Height Regulations) no definition is given; some Health & Safety professionals have had many sleepless nights trying to think up the most suitable definition. Competency is a combination of knowledge, skill and experience. Perhaps the best definition I can give is "someone who has an awareness of the limitations of their own experience and knowledge, who has an understanding of current best practice, who has a willingness and ability to supplement existing experience and knowledge, who has access to specific applied knowledge and skills of appropriately qualified specialists. As far as an employer is concerned, it is his obligation to ensure that the person he chooses for a particular job or task has been adequately trained and is 'competent'.

The words 'shall', 'shall not', 'effective' and 'efficient' impose an absolute duty.

'Maintained' means "kept in efficient working order at all times".

Abbreviations

H&S	Health & Safety
HASAWA	Health & Safety at Work etc Act 1974
HSE	Health & Safety Executive
HSC	Health & Safety Commission
PSA	Production Services Association
Regs. (regs.)	Regulations
EHO	Environmental Health Officer
PPE	Personal Protective Equipment
UK CMA	UK Crowd Management Association

A2 FURTHER INFORMATION

Further information can be obtained directly from the Health & Safety Executive (HSE InfoLine: 0541 545500). They have a duty to give advice and information as well as enforce; they also have an Accident Prevention Unit.

There are a number of H&S magazines available on subscription, including the HSE Newsletter.

There are consultants such as Stagesafe, who can provide professional advice and safety awareness training, and companies that provide plant and equipment training.

A lot of good information can be obtained from the various books, leaflets and guides published by HSE Books. Many of the leaflets are free and a free catalogue is also available. These publications can be obtained directly from HSE Books (address below) or most good bookshops. Once you have ordered any publications from HSE Books (including the free leaflets) they will give you a customer number to be quoted with your next order.

The HSE now has its own website. A large amount of useful information can be obtained at this URL, including information on HSE books and videos, the Faxline, HSE Citizens Charter, new and up and coming issues, as well as local HSE contacts. The URL address for the HSE Information Service is: www.open.gov.uk/hse/hsehome.htm

The more times you contact the HSE or HSE Books the better. Each time you call or make an order for publications is recorded, so if push came to shove, this may be used as further evidence in your favour in court to show that you take Health and Safety seriously and had made efforts to at least gain information. When dealing with HSE officers (even on the phone) make a note of the time and date of the call, together with the officer's name and department.

HSE Books,
Customer Services Dept.,
PO Box 1999. SUDBURY, Suffolk CO10 6FS
Tel: 01787 881165 Fax: 01787 313995
Website: www.hsebooks.co.uk

HSE Information Centre (General Enquires)
Broad Land,
Sheffield
S3 7HQ
Tel: 0541 545500
HSE FastFax Service

A number of selected Health & Safety publications can be obtained from the HSE FastFax service. Simply dial 0839 06 06 06 from a telephone connected to a fax machine and follow the instructions. This service is available 24 hours a day.

Inland Revenue

Your local tax office or enquiry office can help with queries on employment status as well as the obvious tax and national insurance information. Like the HSE they produce numerous free leaflets and information booklets. The Inland Revenue website containing large amounts of useful information can be found at: www.inlandrevenue.gov.uk

A3 FURTHER READING

The Event Safety Guide, A Guide to Health, Safety and Welfare at Music and Similar Events. HSE 1999

Management of Health and Safety at Work. (Management of Health and Safety Work Regulations 1999) ACOP 1999. HSE

Workplace Health, Safety and Welfare. (Workplace (Health, Safety and Welfare) Regulations 1992) ACOP. HSE 1992

Safe Use of Work Equipment. (Provision and Use of Work Equipment Regulations 1998) ACOP and Guidance. HSE 1998

Manual Handling (Manual Handling Operations Regulations 1992) Guidance. HSE 1992

Personal Protective Equipment at Work (Personal Protective Equipment Regulations 1992) Guidance. HSE 1992

Work with Display Screen Equipment (Health and Safety (Display Screen Equipment) Regulations 1999 Guidance on Regulations. HSE 2003

Fire Safety. An Employer's Guide. Home Office 1999 (From HSE Books)

Electricity at Work. Safe Working Practices. HSE 2003

Maintaining Portable and Transportable Electrical Equipment. HSE 1994

Memorandum of Guidance on the Electricity at Work Regulations *1989*. HSE 2000

Safety in Electrical Testing - General Guidance. HSE 2002

Safety in Electrical Testing - Servicing and Repair of Audio, TV and Computer Equipment. HSE 2002

Safety in Electrical Testing: Switch Gear and Control Gear. HSE 2002

Safety in Electrical Testing - Servicing and Repair of Domestic Appliances. HSE 2002

First Aid at Work (The Health and Safety (First-Aid) Regulations 1981) ACOP and Guidance. HSE 1997

Safety Signs and Signals (The Health and Safety (Safety Signs and Signals) Regulations 1996. HSE 1996

A Guide to the Health and Safety (Consultation with Employees) Regulations 1996. HSE 1996

Control of Substances Hazardous to Health Regulations 2002 (ACOP), Control of Carcinogenic Substances (Carcinogens ACOP) and Control of Biological Agents (Biological Agents ACOP). — all published as one document. HSE 2002

Dangerous Substances and Explosive Atmosphere Regulations 2002. ACOP and Guidance. HSE 2003

A Guide to the Reporting of Injuries, Diseases and Dangerous Occurrences Regulations 1995 (RIDDOR). HSE 1995

Safe Use of Lifting Equipment (Lifting Operations and Lifting Equipment Regulations 1998). HSE 1998

Reducing Noise at Work. Guidance on the Noise at Work Regulations 1989. HSE 1998

Construction (Head Protection) Regulations 1989 Guidance. HSE 1990

Health and Safety in Kitchens and Food Preparation Areas. HSE 1991

Safety in Working with Lift Trucks. HSE 1992

Working Platforms on Fork Lift Trucks. HSE 2000

Rider-operated Lift Trucks - Operator Training ACOP and Guidance. HSE 1994

Working Together on Firework Displays. A Guide to Safety for Firework Display Organisers and Operators. HSE 1995

Managing Crowds Safely. HSE 2001

Research to Develop A Methodology for the Assessment of Risks to Crowd Safety in Public Venues. Parts 1 and 2. HSE 1999

Electrical Safety for Entertainers. HSE 1991

Safety Representatives and Safety Committees. HSE 1996

Workplace Transport Safety. Guidance for Employers. HSE 1995

Driving at Work – Managing Work Related Road Safety. HSE 2003

The Costs of Accidents at Work. HSE 1997

Giving your own Firework Display. How to run and fire it safely. HSE 1995

Manual Handling. Solutions you can handle. HSE 1994

Successful Health and Safety Management. HSE 1997

Essentials of Health and Safety at Work. HSE 1999

Health and Safety in Construction. HSE 1996

Managing Contractors. A Guide for Employers. HSE 1997

Safe Use of Ladders, Step Ladders and Trestles. HSE 1993

Electrical Safety at Places of Entertainment. HSE 1991

Working at Heights in the Broadcasting and Entertainment Industries. HSE 1998

General Access Scaffolds and Ladders. HSE 1997

Smoke and Vapour Effects used in Entertainment. HSE 1996

Health and Safety (Miscellaneous Amendments) Regulations 2002. HSE

Non HSE-Publications

An Introduction to Health and Safety Management in the Live Music Industry. Chris Hannam. Production Services Association 1997

Temporary demountable structures. Guidance on Design, Procurement and Use. Second Edition. The Institution of Structural Engineers 1999

Guide to Fire Precautions in Existing Places of Entertainment and Like Places. Home Office and Scottish Office and Health Department. HMSO 1990

Fire Precautions in the Workplace. Information to Employers about the Fire Precautions (Workplace) Regulations 1997 and Fire Precautions (Workplace) (Amendment) Regulations 1999. Home Office and Scottish Office. From HMSO

Guide To Fire Precautions In Existing Places Of Work That Require A Fire Certificate. Home Office and Scottish Office. HMSO

Code of Practice for Design and Instillation of Temporary Distribution Systems Delivering A.C. Electrical Supplies for Lighting, Technical Services and Other Entertainment-related Purposes. (B.S. 7909: 1998) British Standards Institute.

Requirements for Electrical Installation. Institute of Electrical Engineers Wiring Regulations. 16th. Edition. (BS 7671: 2002) British Standards Institute and Institute of Electrical Engineers.

Lifting Equipment for Performance, Broadcast and Similar Applications - Part 1: Specification for the Design and Manufacture of the above Stage Equipment (excluding Trusses and Towers). (B.S. 7905 - 1: 2002) British Standards Institute.

Lifting Equipment for performance, broadcast and similar applications - Part 2: Specification for the Design and Manufacture of Aluminium and Steel Towers and Trusses. (B.S. 7905 - 2: 2002) British Standards Institute.

Lifting equipment for performance, broadcast and similar applications - Part 1: Code of Practice for the Use of the above Stage Equipment (excluding Trusses and Towers). (B.S. 7906 - 1) British Standards Institute 2004.

Lifting Equipment for Performance, Broadcast and Similar Applications - Part 2: Code of Practice for Use of Aluminium and Steel Trusses and Towers. (B.S. 7906 - 2: 2000) British Standards Institute.

Code of Practice for the use of rope access methods for industrial use. (B.S. 7985: 2002) British Standards Institute.

Code of Practice for Event Stewarding and Crowd Safety Services. (B.S. 8406: 2003) British Standards Institute.

Code of Practice for in service inspection and testing of electrical equipment. Institute of Electrical Engineers.

Environmental Protection Act 1990. Waste Management. The Duty of Care. Code of Practice. Department of Environment, Scottish Office and Welsh Office. 1991 From HMSO.

Dealing with Disaster. Home Office 1992 From HMSO.

Code of Practice for the Safe Use of Laser Systems, Festival and other Entertainment Lighting Equipment in Public Places. Institution of Lighting Engineers 1995

Food Safety. First Principles. Chartered Institute of Environmental Health 1998

A Guide to Working Time Regulations 1998. DTI 1998

PASMA: Prefabricated Scaffolding Manufacturers' Association Ltd. Operators' Code of Practice.

Liquid Petroleum Gas Association. Various Codes of Practice.

Guide for the Operation of Lasers, Searchlights and Fireworks in United Kingdom Airspace. (CAP 736) Civil Aviation Authority. Directorate of Airspace Policy. 2003